ARDUINO WORKSHOP

ARDUINO WORKSHOP

A Hands-On Introduction with 65 Projects

by John Boxall

no starch press

San Francisco

Printed in USA

Fifth printing

17 16 15   5 6 7 8 9

ISBN-10: 1-59327-448-3
ISBN-13: 978-1-59327-448-1

Publisher: William Pollock
Production Editor: Serena Yang
Cover Illustration: Charlie Wylie
Interior Design: Octopod Studios
Developmental Editor: William Pollock
Technical Reviewer: Marc Alexander
Copyeditor: Lisa Theobald
Compositor: Susan Glinert Stevens
Proofreader: Emelie Battaglia

Circuit diagrams made using Fritzing (*http://fritzing.org/*)

For information on distribution, translations, or bulk sales, please contact No Starch Press, Inc. directly:

No Starch Press, Inc.
245 8th Street Street, San Francisco, CA 94103
phone: 415.863.9900; info@nostarch.com; www.nostarch.com

Library of Congress Cataloging-in-Publication Data

```
Boxall, John, 1975-
   Arduino workshop : a hands-on introduction with 65 projects / by John Boxall.
        pages cm
   Includes index.
   ISBN-13: 978-1-59327-448-1
   ISBN-10: 1-59327-448-3
  1.  Arduino (Microcontroller)  I. Title.
   TJ223.P76B68 2013
   629.8'95--dc23
                                              2013008261
```

For the two people who have always believed in me:
my mother and my dearest Kathleen

BRIEF CONTENTS

Acknowledgments xix

Chapter 1: Getting Started 1

Chapter 2: Exploring the Arduino Board and the IDE 19

Chapter 3: First Steps 33

Chapter 4: Building Blocks 55

Chapter 5: Working with Functions 95

Chapter 6: Numbers, Variables, and Arithmetic 111

Chapter 7: Liquid Crystal Displays 147

Chapter 8: Expanding Your Arduino 161

Chapter 9: Numeric Keypads 187

Chapter 10: Accepting User Input with Touchscreens 195

Chapter 11: Meet the Arduino Family 207

Chapter 12: Motors and Movement 225

Chapter 13: Using GPS with Your Arduino 257

Chapter 14: Wireless Data 271

Chapter 15: Infrared Remote Control 285

Chapter 16: Reading RFID Tags 295

Chapter 17: Data Buses 307

Chapter 18: Real-time Clocks 321

Chapter 19: The Internet 337

Chapter 20: Cellular Communications 349

Index 365

CONTENTS IN DETAIL

ACKNOWLEDGMENTS   xix

1
GETTING STARTED   1

The Possibilities Are Endless 2
Strength in Numbers. 6
Parts and Accessories. 6
Required Software 7
    Mac OS X 7
    Windows XP and Later 11
    Ubuntu Linux 9.04 and Later 15
Safety. 18
Looking Ahead 18

2
EXPLORING THE ARDUINO BOARD AND THE IDE   19

The Arduino Board. 19
Taking a Look Around the IDE 25
    The Command Area 25
    The Text Area 26
    The Message Window Area 26
Creating Your First Sketch in the IDE. 27
    Comments 27
    The Setup Function 28
    Controlling the Hardware. 28
    The Loop Function 28
    Verifying Your Sketch 30
    Uploading and Running Your Sketch 31
    Modifying Your Sketch. 31
Looking Ahead 31

3
FIRST STEPS   33

Planning Your Projects 34
About Electricity. 34
    Current 34
    Voltage 35
    Power 35
Electronic Components 35
    The Resistor 35
    The Light-Emitting Diode. 39
    The Solderless Breadboard. 41
Project #1: Creating a Blinking LED Wave 43
    The Algorithm 43
    The Hardware 43

The Sketch 43
The Schematic 44
Running the Sketch 45
Using Variables 45
Project #2: Repeating with for Loops **46**
Varying LED Brightness with Pulse-Width Modulation 47
Project #3: Demonstrating PWM **49**
More Electric Components. 49
The Transistor 50
The Rectifier Diode 50
The Relay. 51
Higher-Voltage Circuits. 52
Looking Ahead 53

4
BUILDING BLOCKS   55

Using Schematic Diagrams 56
Identifying Components 56
Wires in Schematics 58
Dissecting a Schematic 59
The Capacitor 60
Measuring the Capacity of a Capacitor 60
Reading Capacitor Values 61
Types of Capacitors. 61
Digital Inputs 63
Project #4: Demonstrating a Digital Input. **65**
The Algorithm 65
The Hardware 65
The Schematic 65
The Sketch 69
Modifying Your Sketch. 70
Understanding the Sketch. 70
Creating Constants with #define 70
Reading Digital Input Pins 70
Making Decisions with if 71
Making More Decisions with if-then-else. 71
Boolean Variables 72
Comparison Operators 72
Making Two or More Comparisons. 73
Project #5: Controlling Traffic **74**
The Goal 74
The Algorithm. 74
The Hardware 75
The Schematic 75
The Sketch 76
Running the Sketch 79
Analog vs. Digital Signals. 79
Project #6: Creating a Single-Cell Battery Tester **80**
The Goal 81
The Algorithm 81
The Hardware 81

The Schematic 81
The Sketch 82
Doing Arithmetic with an Arduino 83
Float Variables 84
Comparison Operators for Calculations 84
Improving Analog Measurement Precision with a Reference Voltage 84
Using an External Reference Voltage 85
Using the Internal Reference Voltage 86
The Variable Resistor 86
Piezoelectric Buzzers 87
Piezo Schematic 88
Project #7: Trying Out a Piezo Buzzer **88**
Project #8: Creating a Quick-Read Thermometer **90**
The Goal 90
The Hardware 90
The Schematic 91
The Sketch 91
Hacking the Sketch 93
Looking Ahead 93

5
WORKING WITH FUNCTIONS   **95**
Project #9: Creating a Function to Repeat an Action **96**
Project #10: Creating a Function to Set the Number of Blinks **97**
Creating a Function to Return a Value 98
Project #11: Creating a Quick-Read Thermometer That Blinks the Temperature **98**
The Hardware 99
The Schematic 99
The Sketch 100
Displaying Data from the Arduino in the Serial Monitor 101
The Serial Monitor 102
Project #12: Displaying the Temperature in the Serial Monitor **103**
Debugging with the Serial Monitor 105
Making Decisions with while Statements 105
do-while 105
Sending Data from the Serial Monitor to the Arduino 106
Project #13: Multiplying a Number by Two **106**
long Variables 107
Project #14: Using long Variables **107**
Looking Ahead 109

6
NUMBERS, VARIABLES, AND ARITHMETIC   **111**
Generating Random Numbers 112
Using Ambient Current to Generate a Random Number 112
Project #15: Creating an Electronic Die **113**
The Hardware 114
The Schematic 114
The Sketch 115
Modifying the Sketch 116

A Quick Course in Binary 116
        Byte Variables 117
Increasing Digital Outputs with Shift Registers 118
Project #16: Creating an LED Binary Number Display. 119
        The Hardware 119
        Connecting the 74HC595 119
        The Sketch 121
Project #17: Making a Binary Quiz Game 122
        The Algorithm. 122
        The Sketch 122
Arrays 124
        Defining an Array 124
        Referring to Values in an Array 125
        Writing to and Reading from Arrays 125
Seven-Segment LED Displays 126
        Controlling the LED 127
Project #18: Creating a Single-Digit Display. 129
        The Hardware 129
        The Schematic 129
        The Sketch 130
        Displaying Double Digits 131
Project #19: Controlling Two Seven-Segment LED Display Modules 131
        The Hardware 131
        The Schematic 132
        Modulo 133
Project #20: Creating a Digital Thermometer 134
        The Hardware 134
        The Sketch 134
LED Matrix Display Modules 135
        The LED Matrix Schematic 136
        Making the Connections 137
Bitwise Arithmetic. 139
        The Bitwise AND Operator. 139
        The Bitwise OR Operator 139
        The Bitwise XOR Operator 140
        The Bitwise NOT Operator. 140
        Bitshift Left and Right 140
Project #21: Creating an LED Matrix 141
Project #22: Creating Images on an LED Matrix 142
Project #23: Displaying an Image on an LED Matrix. 144
Project #24: Animating an LED Matrix. 145
        The Sketch 145
Looking Ahead 146

**7
LIQUID CRYSTAL DISPLAYS   147**
Character LCD Modules 148
        Using a Character LCD in a Sketch 149
        Displaying Text 150
        Displaying Variables or Numbers 151

Project #25: Defining Custom Characters **152**
Graphic LCD Modules 153
    Connecting the Graphic LCD 154
    Using the LCD. 155
    Controlling the Display 155
Project #26: Seeing the Text Functions in Action **155**
    Creating More Complex Display Effects 156
Project #27: Creating a Temperature History Monitor **157**
    The Algorithm 158
    The Hardware 158
    The Sketch 158
    The Result. 160
    Modifying the Sketch. 160
Looking Ahead 160

8
EXPANDING YOUR ARDUINO   161

Shields 162
ProtoShields 164
Project #28: Creating a Custom Shield with Eight LEDs **165**
    The Hardware 165
    The Schematic 165
    The Layout of the ProtoShield Board 166
    The Design 166
    Soldering the Components 167
    Modifying the Custom Shield 169
Expanding Sketches with Libraries 169
    Importing a Shield's Libraries 169
MicroSD Memory Cards 173
    Testing Your MicroSD Card 174
Project #29: Writing Data to the Memory Card **175**
Project #30: Creating a Temperature-Logging Device **177**
    The Hardware 177
    The Sketch 177
Timing Applications with millis() and micros(). 179
Project #31: Creating a Stopwatch **181**
    The Hardware 181
    The Schematic 181
    The Sketch 182
Interrupts. 184
    Interrupt Modes. 184
    Configuring Interrupts 185
    Activating or Deactivating Interrupts 185
Project #32: Using Interrupts **185**
    The Sketch 185
Looking Ahead 186

9
NUMERIC KEYPADS 187

Using a Numeric Keypad 187
    Wiring a Keypad 188
    Programming for the Keypad 189
    Testing the Sketch 189
Making Decisions with switch-case 190
Project #33: Creating a Keypad-Controlled Lock **190**
    The Sketch 191
    How It Works 192
    Testing the Sketch 193
Looking Ahead 193

10
ACCEPTING USER INPUT WITH TOUCHSCREENS 195

Touchscreens 195
    Connecting the Touchscreen 196
Project #34: Addressing Areas on the Touchscreen **197**
    The Hardware 197
    The Sketch 197
    Testing the Sketch 198
    Mapping the Touchscreen 199
Project #35: Creating a Two-Zone On/Off Touch Switch **200**
    The Sketch 200
    How It Works 202
    Testing the Sketch 202
Project #36: Creating a Three-Zone Touch Switch **202**
    The Touchscreen Map 203
    The Sketch 203
    How It Works 205
Looking Ahead 205

11
MEET THE ARDUINO FAMILY 207

Project #37: Creating Your Own Breadboard Arduino **208**
    The Hardware 208
    The Schematic 211
    Running a Test Sketch 214
The Many Arduino Boards 217
    Arduino Uno. 219
    Freetronics Eleven 219
    The Freeduino 220
    The Pro Trinket 220
    The Arduino Nano 221
    The Arduino LilyPad. 221
    The Arduino Mega 2560 222
    The Freetronics EtherMega 222
    The Arduino Due. 223
Looking Ahead 224

12
MOTORS AND MOVEMENT   225

Making Small Motions with Servos 225
    Selecting a Servo 226
    Connecting a Servo 227
    Putting a Servo to Work 227
Project #38: Building an Analog Thermometer **228**
    The Hardware 228
    The Schematic 229
    The Sketch 229
Using Electric Motors 231
    The TIP120 Darlington Transistor 231
Project #39: Controlling the Motor **232**
    The Hardware 232
    The Schematic 233
    The Sketch 234
Project #40: Building and Controlling a Tank Robot **235**
    The Hardware 235
    The Schematic 238
    The Sketch 240
Sensing Collisions 243
Project #41: Detecting Tank Bot Collisions with a Microswitch **243**
    The Schematic 243
    The Sketch 244
Infrared Distance Sensors 246
    Wiring It Up 247
    Testing the IR Distance Sensor 247
Project #42: Detecting Tank Bot Collisions with IR Distance Sensor **249**
Ultrasonic Distance Sensors 251
    Connecting the Ultrasonic Sensor 252
    Using the Ultrasonic Sensor 252
    Testing the Ultrasonic Distance Sensor 252
Project #43: Detecting Tank Bot Collisions with an Ultrasonic Distance Sensor **254**
    The Sketch 254
Looking Ahead 256

13
USING GPS WITH YOUR ARDUINO   257

What Is GPS? 258
Testing the GPS Shield 259
Project #44: Creating a Simple GPS Receiver **261**
    The Hardware 261
    The Sketch 261
    Displaying the Position on the LCD 262
Project #45: Creating an Accurate GPS-based Clock **263**
    The Hardware 263
    The Sketch 264

Project #46: Recording the Position of a Moving Object over Time 265
    The Hardware 265
    The Sketch 266
    Displaying Locations on a Map 268
Looking Ahead 269

14
WIRELESS DATA   271

Using Low-cost Wireless Modules 271
Project #47: Creating a Wireless Remote Control 272
    The Hardware for the Transmitter Circuit 273
    The Transmitter Schematic 273
    The Hardware for the Receiver Circuit 274
    The Receiver Schematic 274
    The Transmitter Sketch 275
    The Receiver Sketch 276
Using XBee Wireless Data Modules for Greater Range and Faster Speed 277
Project #48: Transmitting Data with an XBee 279
    The Sketch 279
    Setting Up the Computer to Receive Data 279
Project #49: Building a Remote Control Thermometer 281
    The Hardware 281
    The Layout 281
    The Sketch 282
    Operation 283
Looking Ahead 284

15
INFRARED REMOTE CONTROL   285

What Is Infrared? 285
Setting Up for Infrared 286
    The IR Receiver 286
    The Remote Control 287
    A Test Sketch 287
    Testing the Setup 288
Project #50: Creating an IR Remote Control Arduino 289
    The Hardware 289
    The Sketch 289
    Expanding the Sketch 290
Project #51: Creating an IR Remote Control Tank 291
    The Hardware 291
    The Sketch 291
Looking Ahead 293

16
READING RFID TAGS   295

Inside RFID Devices 296
Testing the Hardware 297
    The Schematic 297
    Testing the Schematic 297

Project #52: Creating a Simple RFID Control System **299**
    The Sketch 299
    How It Works 300
Storing Data in the Arduino's Built-in EEPROM 301
    Reading and Writing to the EEPROM 302
Project #53: Creating an RFID Control with "Last Action" Memory **303**
    The Sketch 303
    How It Works 306
Looking Ahead 306

17
DATA BUSES   307

The I^{2}C Bus 308
Project #54: Using an External EEPROM **309**
    The Hardware 309
    The Schematic 310
    The Sketch 311
    The Result 312
Project #55: Using a Port Expander IC **313**
    The Hardware 313
    The Schematic 313
    The Sketch 314
The SPI Bus 315
    Pin Connections 316
    Implementing the SPI 316
    Sending Data to an SPI Device 317
Project #56: Using a Digital Rheostat **318**
    The Hardware 318
    The Schematic 318
    The Sketch 319
Looking Ahead 320

18
REAL-TIME CLOCKS   321

Connecting the RTC Module 322
Project #57: Adding and Displaying Time and Date with an RTC **322**
    The Hardware 322
    The Sketch 323
    How It Works 325
Project #58: Creating a Simple Digital Clock **326**
    The Hardware 326
    The Sketch 327
    How It Works and Results 330
Project #59: Creating an RFID Time-Clock System **330**
    The Hardware 331
    The Sketch 331
    How It Works 335
Looking Ahead 336

19
THE INTERNET   337

What You'll Need 337
Project #60: Building a Remote-Monitoring Station 339
    The Hardware 339
    The Sketch 339
    Troubleshooting 341
    How It Works 342
Project #61: Creating an Arduino Tweeter 343
    The Hardware 343
    The Sketch 343
Controlling Your Arduino from the Web 344
Project #62: Setting Up a Remote Control for Your Arduino 345
    The Hardware 345
    The Sketch 346
    Controlling Your Arduino Remotely 347
Looking Ahead 348

20
CELLULAR COMMUNICATIONS   349

The Hardware 350
    Preparing the Power Shield 351
    Hardware Configuration and Testing 352
    Changing the Operating Frequency 354
Project #63: Building an Arduino Dialer 356
    The Hardware 356
    The Schematic 356
    The Sketch 357
    How It Works 358
Project #64: Building an Arduino Texter 358
    The Sketch 359
    How It Works 359
Project #65: Setting Up an SMS Remote Control 360
    The Hardware 360
    The Schematic 361
    The Sketch 361
    How It Works 363
Looking Ahead 364

INDEX   365

ACKNOWLEDGMENTS

First of all, a huge thank you to the Arduino team: Massimo Banzi, David Cuartielles, Tom Igoe, Gianluca Martino, and David Mellis. Without your vision, thought, and hard work, none of this would have been possible.

Many thanks to my technical reviewer Marc Alexander for his contributions, expertise, suggestions, support, thoughts, and long conversations, and for having the tenacity to follow through with such a large project.

I also want to thank the following organizations for their images and encouragement: adafruit industries, Agilent Technologies, Gravitech, Freetronics, Oomlout, Seeed Studio, Sharp Corporation, and SparkFun. Furthermore, a big thanks to Freetronics for the use of their excellent hardware products. And thank you to all those who have contributed their time making Arduino libraries, which makes life much easier for everyone.

Kudos and thanks to the Fritzing team for their wonderful open source circuit schematic design tool, which I've used throughout this book.

And a thank you to the following people (in no particular order) from whom I've received encouragement, inspiration and support: Iraphne Childs, Limor Fried, Jonathan Oxer, Philip Lindsay, Nicole Kilah, Ken Shirriff, Nathan Kennedy, David Jones, and Nathan Seidle.

Finally, thank you to everyone at No Starch Press, including Sondra Silverhawk for suggesting the book; Serena Yang for her dedicated editing, endless patience, and suggestions; and Bill Pollock for his support and guidance and for convincing me that sometimes there is a better way to explain something.

1

GETTING STARTED

Have you ever looked at some gadget and wondered how it *really* worked? Maybe it was a remote control boat, the system that controls an elevator, a vending machine, or an electronic toy? Or have you wanted to create your own robot or electronic signals for a model railroad, or perhaps you'd like to capture and analyze weather data over time? Where and how do you start?

The Arduino board (shown in Figure 1-1) can help you find some of the answers to the mysteries of electronics in a hands-on way. The original creation of Massimo Banzi and David Cuartielles, the Arduino system offers an inexpensive way to build interactive projects, such as remote-controlled robots, GPS tracking systems, and electronic games.

The Arduino project has grown exponentially since its introduction in 2005. It's now a thriving industry, supported by a community of people united with the common bond of creating something new. You'll find both individuals and groups, ranging from interest groups and clubs to local hackerspaces and educational institutions, all interested in toying with the Arduino.

Figure 1-1: The Arduino board

To get a sense of the variety of Arduino projects in the wild, you can simply search the Internet. You'll find a list of groups offering introductory programs and courses with like-minded, creative people.

The Possibilities Are Endless

A quick scan through this book will show you that you can use the Arduino to do something as simple as blinking a small light, or even something more complicated, such as interacting with a cellular phone—and many different things in between.

For example, have a look at Philip Lindsay's device, shown in Figure 1-2. It can receive text messages from cellular phones and display them on a large sign for use in dance halls. This device uses an Arduino board and a cellular phone shield to receive text messages from other phones (similar to Project 65). The text message is sent to a pair of large, inexpensive dot-matrix displays for everyone to see.

Figure 1-2: SMS (short message service) text marquee

You can purchase large display boards that are easy to interface with an Arduino, so you don't have to make your own display from scratch. (For more information, visit *http://www.labradoc.com/i/follower/ p/project-sms-text-scroller.*)

How about creating a unique marriage proposal? Tyler Cooper wanted an original way to propose to his girlfriend, so he built what he calls a "reverse geocache box"—a small box that contained an engagement ring, as shown in Figure 1-3. When the box was taken to a certain area (measured by the internal GPS), it unlocked to reveal a romantic message and the ring. You can easily reproduce this device using an Arduino board, a GPS receiver, and an LCD module (as used in Chapter 13), with a small servo motor that acts as a latch to keep the box closed until it's in the correct location. The code required to create this is quite simple—something you could create in a few hours. The most time-consuming part is choosing the appropriate box in which to enclose the system. (For more information, visit *http://learn.adafruit.com/reverse-geocache-engagement-box/.*)

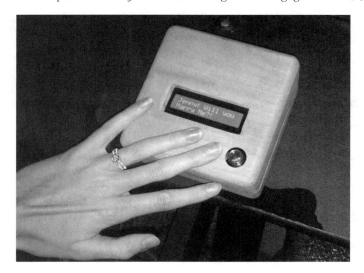

Figure 1-3: Marriage proposal via Arduino

Here's another example. Kurt Schulz was interested in monitoring the battery charge level of his moped. However, after realizing how simple it is to work with Arduino, his project morphed into what he calls the "Scooterputer": a complete moped management system. The Scooterputer can measure the battery voltage, plus it can display the speed, distance traveled, tilt angle, temperature, time, date, GPS position, and more. It also contains a cellular phone shield that can be controlled remotely, allowing remote tracking of the moped and engine shutdown in case it's stolen. The entire system can be controlled with a small touchscreen, shown in

Figure 1-4. Each feature can be considered a simple building block, and anyone could create a similar system in a couple of weekends. (See *http://www.janspace.com/b2evolution/arduino.php/2010/06/26/scooterputer/.*)

Figure 1-4: The Scooterputer display (courtesy of Kurt Schulz)

Then there's John Sarik, who enjoys the popular Sudoku math puzzles; he also likes working with Nixie numeric display tubes. With those two drivers in mind, John created a huge 81-digit Sudoku game computer! The user can play a full 9-by-9 game, with the Arduino in control of the digits and checking for valid entries. Although this project might be considered a more advanced type, it is certainly achievable and the electronics are not complex. The device is quite large and looks great mounted on a wall, as shown in Figure 1-5. (See *http://trashbearlabs.wordpress.com/2010/07/09/nixie-sudoku/.*)

The team at Oomlout even used the Arduino to create a TwypeWriter. They fitted an Arduino board with an Ethernet shield interface connected to the Internet, which searches Twitter for particular keywords. When a keyword is found, the tweet is sent to an electric typewriter for printing. The Arduino board is connected to the typewriter's keyboard circuit, which allows it to emulate a real person typing, as shown in Figure 1-6. (See *http://oomlout.co.uk/blog/twitter-monitoring-typewritter-twypwriter/.*)

These are only a few random examples of what is possible using an Arduino. You can create your own projects without much difficulty—and after you've worked through this book, they are certainly not out of your reach.

Figure 1-5: Nixie tube Sudoku

Figure 1-6: The TwypeWriter

Strength in Numbers

The Arduino platform increases in popularity every day. If you're more of a social learner and enjoy class-oriented situations, search the Web for "Cult of Arduino" to see what people are making and to find Arduino-related groups. Members of Arduino groups introduce the world of Arduino from an artist's perspective. Many group members work to create a small Arduino-compatible board at the same time. These groups can be a lot of fun, introduce you to interesting people, and let you share your Arduino knowledge with others.

Parts and Accessories

As with any other electronic device, the Arduino is available from many retailers that offer a range of products and accessories. When you're shopping, be sure to purchase the original Arduino, not a knock-off, or you run the risk of receiving faulty or poorly performing goods; why risk your project with an inferior board that could end up costing you more in the long run? For a list of Arduino suppliers, visit *http://arduino.cc/en/Main/Buy/*.

Here's a list of current suppliers (in alphabetical order) that I recommend for your purchases of Arduino-related parts and accessories:

- Adafruit Industries (*http://www.adafruit.com/*)
- Altronics (*http://www.altronics.com.au/*)
- DigiKey (*http://www.digikey.com/*)
- Jameco Electronics (*http://www.jameco.com/*)
- Jaycar (*http://www.jaycar.com.au/*)
- Newark (*http://www.newark.com/*)
- nicegear (*http://www.nicegear.co.nz/*)
- Oomlout (*http://www.oomlout.co.uk/*)
- SparkFun Electronics (*http://www.sparkfun.com/*)
- Tronixlabs (*http://tronixlabs.com/*)

As you'll see in this book, I use several Arduino-compatible products from Freetronics (*http://www.freetronics.com/*). However, you will find that all the required parts are quite common and easily available from various resellers.

But don't go shopping yet. Take the time to read the first few chapters to get an idea of what you'll need so that you won't waste money buying unnecessary things immediately.

Required Software

You should be able to program your Arduino with just about any computer using a piece of software called an *integrated development environment (IDE)*. To run this software, your computer should have one of the following operating systems installed:

- Mac OS X or higher
- Windows XP 32- or 64-bit, or higher
- Linux 32- or 64-bit (Ubuntu or similar)

Now is a good time to download and install the IDE, so jump to the heading that matches your operating system and follow the instructions. Make sure you have or buy the matching USB cable for your Arduino from the supplier as well. Even if you don't have your Arduino board yet, you can still download and explore the IDE. Because the IDE version number can change quite rapidly, the number in this book may not match the current version, but the instructions should still work.

Mac OS X

In this section, you'll find instructions for downloading and configuring the Arduino IDE in Mac OS X.

Installing the IDE

To install the IDE on your Mac, follow these instructions:

1. Using a web browser such as Safari, visit the software download page located at *http://arduino.cc/en/Main/Software/*, as shown in Figure 1-7.

Figure 1-7: The IDE download page in Safari

2. Click the **Mac OS X** link. The file will start downloading, and it will appear in the Downloads window shown in Figure 1-8.

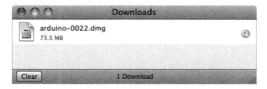

Figure 1-8: File download is complete.

3. Once it's finished downloading, double-click the file to start the installation process. You may be asked if you want to open the file; if so, click **Open**. After a moment the downloaded file will be converted into an Application.

4. Drag the Arduino icon over the Applications folder and release the mouse button. A temporary status window will appear as the file is copied.

5. Now connect your Arduino to your Mac with the USB cable. After a moment, the dialog shown in Figure 1-9 will appear.

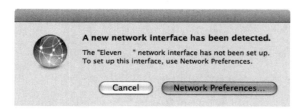

Figure 1-9: A new Arduino board is detected. Your dialog may read Uno *instead of* Eleven.

6. Click **Network Preferences...**, and then click **Apply** in the Network box. You can ignore the "not configured" status message.

Setting Up the IDE

Once you have downloaded the IDE, use the following instructions to open and configure the IDE:

1. Open the Applications folder in Finder (shown in Figure 1-10) and double-click the Arduino icon.

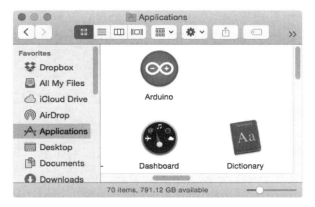

Figure 1-10: Your Applications folder

2. A window may appear warning you about opening a web app. If it does, click **Open** to continue. You will then be presented with the IDE, as shown in Figure 1-11.

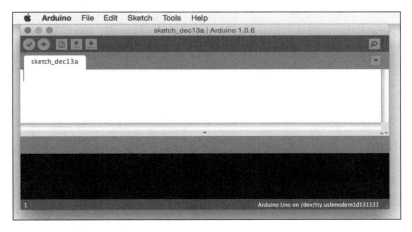

Figure 1-11: The IDE in Mac OS X

3. You're almost there—just two more things to do before your Arduino IDE is ready to use. First, you need to tell the IDE which type of socket the Arduino is connected to. Select **Tools ▶ Serial Port** and select the **/dev/tty.usbmodem1d11** option, as shown in Figure 1-12.

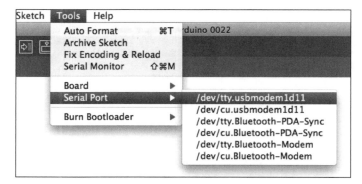

Figure 1-12: Selecting the USB port

4. The final step is to tell the IDE which Arduino board you have connected. This is crucial, since Arduino boards do differ. For example, if you have the most common board, the Uno, then select **Tools ▶ Board ▶ Arduino Uno**, as shown in Figure 1-13. The differences in Arduino boards are explained in more detail in Chapter 11.

Now your hardware and software are ready to work for you. Next, move on to "Safety" on page 18.

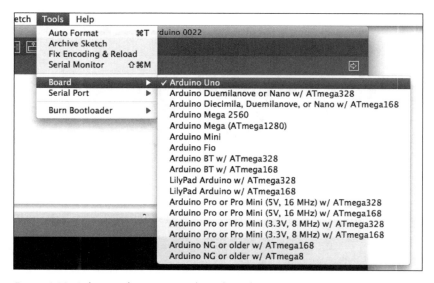

Figure 1-13: Selecting the correct Arduino board

Windows XP and Later

In this section, you'll find instructions for downloading the IDE, installing drivers, and configuring the IDE in Windows.

Installing the IDE

To install the Arduino IDE for Windows, follow these instructions:

1. Using a web browser such as Firefox, visit the software download page located at *http://arduino.cc/en/Main/Software/*, as shown in Figure 1-14.

Figure 1-14: The IDE download page in Windows Firefox

2. Click the **Windows** link, and the dialog shown in Figure 1-15 will appear. Select **Open with Windows Explorer**, and then click **OK**. The file will start to download, as shown in Figure 1-16.

Figure 1-15: Downloading the file

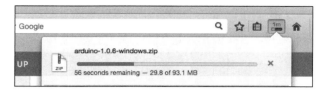

Figure 1-16: Firefox shows the progress of your download.

3. Once the download is complete, double-click the file, and the window shown in Figure 1-17 will appear.

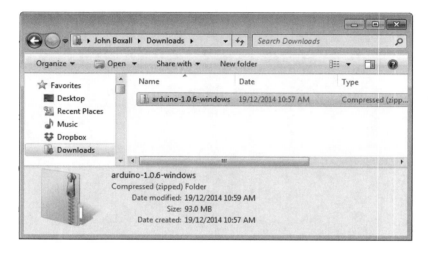

Figure 1-17: The IDE package

4. Copy the folder named *arduino-1.0.6-windows* (or something similar) to the location where you store your applications. Once the copying is finished, locate the folder and open it to reveal the Arduino application icon, as shown in Figure 1-18. You may wish to copy the icon and place a shortcut on the desktop for easier access in the future.

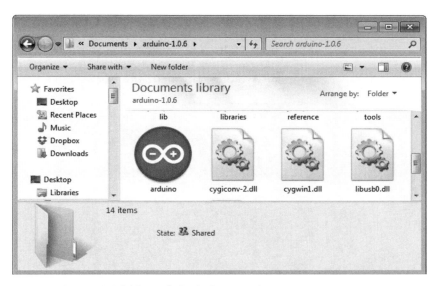

Figure 1-18: Your IDE folder with the Arduino application icon

Installing Drivers

The next task is to install the drivers for your Arduino board's USB interface.

1. Connect your Arduino to your PC with the USB cable. After a few moments an error message will be displayed, which will say something like "Device driver software not successfully installed." Just close that dialog or balloon.

2. Navigate to the Windows Control Panel. Open the Device Manager and scroll down until you see the Arduino, as shown in Figure 1-19.

Figure 1-19: The Device Manager

3. Right-click **Arduino Uno** under Other Devices and select **Update Driver Software**. Then, select the **Browse my computer for driver software** option that appears in the next dialog. Another Browse For Folder dialog will appear; click **Browse**, and navigate to the *drivers* folder in the newly installed Arduino software folder (shown in Figure 1-20). Click **OK**.

Figure 1-20: Locating the drivers *folder*

4. Click **Next** in the dialog that follows. Windows may present a message stating that it "cannot verify the publisher of the driver software." Click **Install this software anyway**. After a short wait, Windows will tell you that the driver is installed and the COM port number the Arduino is connected to, as shown in Figure 1-21.

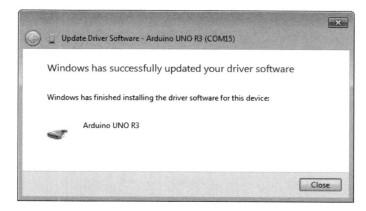

Figure 1-21: The drivers have been updated successfully.

Setting Up the IDE

Okay, we're almost there—just two more things to do to finish setting up the IDE.

1. Open the Arduino IDE. You need to tell the IDE which type of socket the Arduino is connected to by selecting **Tools ▶ Serial Port** and selecting the COM port number that appeared in the Update Driver Software window.

2. The final step is to tell the IDE which Arduino board we have connected. This is crucial, as the Arduino boards do differ. For example, if you have the Uno, select **Tools ▶ Board ▶ Arduino Uno**. The differences in Arduino boards are explained in more detail in Chapter 11.

Now that your Arduino IDE is set up, you can move on to "Safety" on page 18.

Ubuntu Linux 9.04 and Later

If you are running Ubuntu Linux, here are instructions for downloading and setting up the Arduino IDE.

Installing the IDE

Use the following instructions to install the IDE:

1. Using a web browser such as Firefox, visit the software download page located at *http://arduino.cc/en/Main/Software/*, as shown in Figure 1-22.

Figure 1-22: The IDE download page in Ubuntu Firefox

2. In the IDE download page, locate the latest stable version of the IDE, which is currently 1.0.6. Then click the Linux **32-bit** or **64-bit** link, depending on your system. When the dialog in Figure 1-23 appears, select **Open with Archive Manager** and click **OK**.

Figure 1-23: Downloading the file

3. After the file has downloaded, it will be displayed in the Archive Manager, as shown in Figure 1-24. Copy the *arduino-1.0.6* folder (or something similar) to your usual application or Home folder.

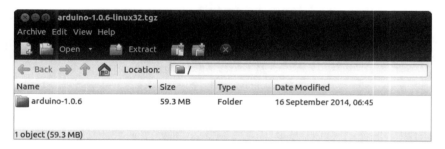

Figure 1-24: The IDE package

Setting Up the IDE

Next, you'll configure the IDE.

1. Connect your Arduino to your PC with the USB cable. At this point you want to run the Arduino IDE, so locate the *arduino-1.0.6* folder you copied earlier and double-click the *arduino* file that's selected in Figure 1-25.

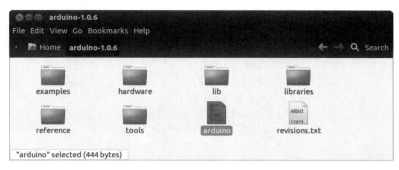

Figure 1-25: Your Arduino IDE folder with the arduino file selected

2. If the dialog shown in Figure 1-26 appears, click **Run**, and you will be presented with the IDE, as shown in Figure 1-27.

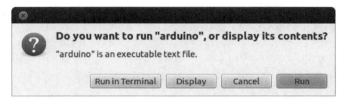

Figure 1-26: Granting permission to run the IDE

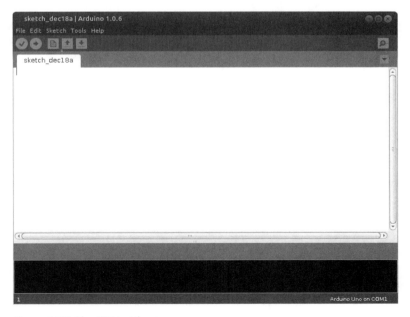

Figure 1-27: The IDE in Ubuntu

3. Now that the IDE is running, we need to tell it which type of socket the Arduino is connected to. Select **Tools ▸ Serial Port** and select the **/dev/ttyACM*x*** port, where *x* is a single digit (there should be only one port with a name like this).

4. Next, tell the IDE which Arduino you have connected. This is crucial, as Arduino boards do differ. For example, if you have the Uno, select **Tools ▸ Board ▸ Arduino Uno**. The differences in Arduino boards are explained in more detail in Chapter 11.

Now your hardware and software are ready to work for you.

Safety

As with any hobby or craft, it's up to you to take care of yourself and those around you. As you'll see in this book, I discuss working with basic hand tools, battery-powered electrical devices, sharp knives, and cutters—and sometimes soldering irons. At no point in your projects should you work with the mains current. Leave that to a licensed electrician who is trained for such work. Remember that contacting the mains current will kill you.

Looking Ahead

You're about to embark on a fun and interesting journey, and you'll be creating things you may never have thought possible. You'll find 65 Arduino projects in this book, ranging from the very simple to the relatively complex. All are designed to help you learn and make something useful. So let's go!

2

EXPLORING THE ARDUINO BOARD AND THE IDE

In this chapter you'll explore the Arduino board as well as the IDE software that you'll use to create and upload Arduino *sketches* (Arduino's name for its programs) to the Arduino board itself. You'll learn the basic framework of a sketch and some basic functions that you can implement in a sketch, and you'll create and upload your first sketch.

The Arduino Board

What exactly is Arduino? According to the Arduino website (*http://www.arduino.cc/*), it is

> an open-source electronics prototyping platform based on flexible, easy-to-use hardware and software. It's intended for artists, designers, hobbyists, and anyone interested in creating interactive objects or environments.

In simple terms, the Arduino is a tiny computer system that can be programmed with your instructions to interact with various forms of input and output. The current Arduino board model, the Uno, is quite small in size compared to the average human hand, as you can see in Figure 2-1.

Figure 2-1: An Arduino Uno is quite small.

Although it might not look like much to the new observer, the Arduino system allows you to create devices that can interact with the world around you. By using an almost unlimited range of input and output devices, sensors, indicators, displays, motors, and more, you can program the exact interactions required to create a functional device. For example, artists have created installations with patterns of blinking lights that respond to the movements of passers-by, high school students have built autonomous robots that can detect an open flame and extinguish it, and geographers have designed systems that monitor temperature and humidity and transmit this data back to their offices via text message. In fact, you'll find an almost infinite number of examples with a quick search on the Internet.

Now let's move on and explore our Arduino Uno *hardware* (in other words, the "physical part") in more detail and see what we have. Don't worry too much about understanding what you see here, because all these things will be discussed in greater detail in later chapters.

Let's take a quick tour of the Uno. Starting at the left side of the board, you'll see two connectors, as shown in Figure 2-2.

Figure 2-2: The USB and power connectors

On the far left is the Universal Serial Bus (USB) connector. This connects the board to your computer for three reasons: to supply power to the board, to upload your instructions to the Arduino, and to send data to and receive it from a computer. On the right is the power connector. Through this connector, you can power the Arduino with a standard mains power adapter.

At the lower middle is the heart of the board: the microcontroller, as shown in Figure 2-3.

Figure 2-3: The microcontroller

The *microcontroller* is the "brains" of the Arduino. It is a tiny computer that contains a processor to execute instructions, includes various types of memory to hold data and instructions from our sketches, and provides various avenues of sending and receiving data. Just below the microcontroller are two rows of small sockets, as shown in Figure 2-4.

Figure 2-4: The power and analog sockets

The first row offers power connections and the ability to use an external RESET button. The second row offers six analog inputs that are used to measure electrical signals that vary in voltage. Furthermore, pins A4 and A5 can also be used for sending data to and receiving it from other devices. Along the top of the board are two more rows of sockets, as shown in Figure 2-5.

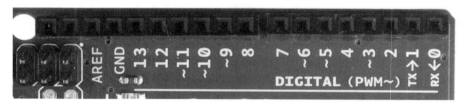

Figure 2-5: The digital input/output pins

Sockets (or pins) numbered 0 to 13 are digital input/output (I/O) pins. They can either detect whether or not an electrical signal is present or generate a signal on command. Pins 0 and 1 are also known as the *serial port*, which is used to send and receive data to other devices, such as a computer via the USB connector circuitry. The pins labeled with a tilde (~) can also generate a varying electrical signal, which can be useful for such things as creating lighting effects or controlling electric motors.

Next are some very useful devices called *light-emitting diodes (LEDs)*; these very tiny devices light up when a current passes through them. The Arduino board has four LEDs: one on the far right labeled ON, which indicates when the board has power, and three in another group, as shown in Figure 2-6.

The LEDs labeled *TX* and *RX* light up when data is being transmitted or received between the Arduino and attached devices via the serial port and USB. The *L* LED is for your own use (it is connected to the digital I/O pin number 13). The little black square part to the left of the LEDs is a tiny microcontroller that controls the USB interface that allows your Arduino to send data to and receive it from a computer, but you don't generally have to concern yourself with it.

Figure 2-6: The onboard LEDs

And, finally, the RESET button is shown in Figure 2-7.

Figure 2-7: The RESET button

As with a normal computer, sometimes things can go wrong with the Arduino, and when all else fails, you might need to reset the system and restart your Arduino. This simple RESET button on the board (Figure 2-7) is used to restart the system to resolve these problems.

One of the great advantages of the Arduino system is its ease of expandability—that is, it's easy to add more hardware functions. The two rows of sockets along each side of the Arduino allow the connection of a *shield*, another circuit board with pins that allow it to plug into the Arduino. For example, the shield shown in Figure 2-8 contains an Ethernet interface that allows the Arduino to communicate over networks and the Internet, with plenty of space for custom circuitry.

Notice how the Ethernet shield also has rows of sockets. These enable you to insert one or more shields on top. For example, Figure 2-9 shows that another shield with a large numeric display, temperature sensor, extra data storage space, and a large LED has been inserted.

Note that you do need to remember which shield uses which individual inputs and outputs to ensure that "clashes" do not occur. You can also purchase completely blank shields that allow you to add your own circuitry. This will be explained further in Chapter 8.

Figure 2-8: Arduino Ethernet interface shield

Figure 2-9: Numeric display and temperature shield

The companion to the Arduino hardware is the *software*, a collection of instructions that tell the hardware what to do and how to do it. Two types of software can be used: The first is the integrated development environment (IDE), which is discussed in this chapter, and the second is the Arduino sketch you create yourself.

The IDE software is installed on your personal computer and is used to compose and send sketches to the Arduino board.

Taking a Look Around the IDE

As shown in Figure 2-10, the Arduino IDE resembles a simple word processor. The IDE is divided into three main areas: the command area, the text area, and the message window area.

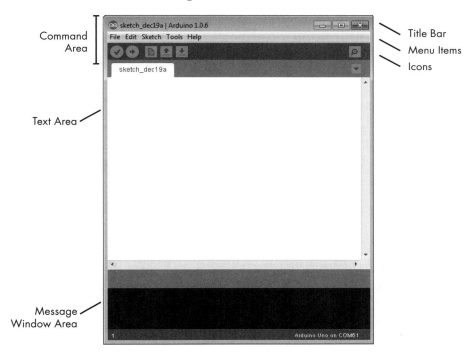

Figure 2-10: The Arduino IDE

The Command Area

The command area is shown at the top of Figure 2-10 and includes the title bar, menu items, and icons. The title bar displays the sketch's filename (*sketch_jun28a*), as well as the version of the IDE (*Arduino 1.0.5*). Below this is a series of menu items (File, Edit, Sketch, Tools, and Help) and icons, as described next.

Menu Items

As with any word processor or text editor, you can click one of the menu items to display its various options.

File   Contains options to save, load, and print sketches; a thorough set of example sketches to open; as well as the **Preferences** submenu

Edit   Contains the usual copy, paste, and search functions common to any word processor

Sketch   Contains the function to verify your sketch before uploading to a board, and some sketch folder and import options

Tools   Contains a variety of functions as well as the commands to select the Arduino board type and USB port

Help   Contains links to various topics of interest and the version of the IDE

The Icons

Below the menu toolbar are six icons. Mouse over each icon to display its name. The icons, from left to right, are as follows:

Verify   Click this to check that the Arduino sketch is valid and doesn't contain any programming mistakes.

Upload   Click this to verify and then upload your sketch to the Arduino board.

New   Click this to open a new blank sketch in a new window.

Open   Click this to open a saved sketch.

Save   Click this to save the open sketch. If the sketch doesn't have a name, you will be prompted to create one.

Serial Monitor   Click this to open a new window for use in sending and receiving data between your Arduino and the IDE.

The Text Area

The text area is shown in the middle of Figure 2-10; this is where you'll create your sketches. The name of the current sketch is displayed in the tab at the upper left of the text area. (The default name is the current date.) You'll enter the contents of your sketch here as you would in any text editor.

The Message Window Area

The message window area is shown at the bottom of Figure 2-10. Messages from the IDE appear in the black area. The messages you see will vary and will include messages about verifying sketches, status updates, and so on.

At the bottom right of the message area, you should see the name of your Arduino board type as well as its connected USB port—*Arduino Uno on COM4* in this case.

Creating Your First Sketch in the IDE

An Arduino sketch is a set of instructions that you create to accomplish a particular task; in other words, a sketch is a *program*. In this section you'll create and upload a simple sketch that will cause the Arduino's LED (shown in Figure 2-11) to blink repeatedly, by turning it on and then off for 1 second intervals.

Figure 2-11: The LED on the Arduino board, next to the capital L

NOTE *Don't worry too much about the specific commands in the sketch we're creating here. The goal is to show you how easy it is to get the Arduino to do something so that you'll keep reading when you get to the harder stuff.*

To begin, connect your Arduino to the computer with the USB cable. Then open the IDE, choose **Tools ▸ Serial Port**, and make sure the USB port is selected. This ensures that the Arduino board is properly connected.

Comments

First, enter a comment as a reminder of what your sketch will be used for. A *comment* is a note of any length in a sketch, written for the user's benefit. Comments in sketches are useful for adding notes to yourself or others, for entering instructions, or for noting miscellaneous details. When programming your Arduino (creating sketches), it's a good idea to add comments regarding your intentions; these comments can prove useful later when you're revisiting a sketch.

To add a comment on a single line, enter two forward slashes and then the comment, like this:

```
// Blink LED sketch by Mary Smith, created 07/01/13
```

The two forward slashes tell the IDE to ignore the text that follows when verifying a sketch. (As mentioned earlier, when you verify a sketch, you're asking the IDE to check that everything is written properly with no errors.)

To enter a comment that spans two or more lines, enter the characters /* on a line before the comment, and then end the comment with the characters */ on the following line, like this:

```
/*
Arduino Blink LED Sketch
by Mary Smith, created 07/01/13
*/
```

As with the two forward slashes that precede a single line comment, the /* and */ tell the IDE to ignore the text that they bracket.

Enter a comment describing your Arduino sketch using one of these methods, and then save your sketch by choosing **File ▸ Save As**. Enter a short name for your sketch (such as *blinky*), and then click **OK**.

The default filename extension for Arduino sketches is *.ino*, and the IDE should add this automatically. The name for your sketch should be, in this case, *blinky.ino*, and you should be able to see it in your Sketchbook.

The Setup Function

The next stage in creating any sketch is to add the void setup() function. This function contains a set of instructions for the Arduino to execute once only, each time it is reset or turned on. To create the setup function, add the following lines to your sketch, after the comments:

```
void setup()
{

}
```

Controlling the Hardware

Our program will blink the user LED on the Arduino. The user LED is connected to the Arduino's digital pin 13. A digital pin can either detect an electrical signal or generate one on command. In this project, we'll generate an electrical signal that will light the LED. This may seem a little complicated, but you'll learn more about digital pins in future chapters. For now, just continue with creating the sketch.

Enter the following into your sketch between the braces ({ and }):

```
pinMode(13, OUTPUT); // set digital pin 13 to output
```

The number 13 in the listing represents the digital pin you're addressing. You're setting this pin to OUTPUT, which means it will generate (output) an electrical signal. If you wanted it to detect an incoming electrical signal, then you would use INPUT instead. Notice that the function pinMode() ends with a semicolon (;). Every function in your Arduino sketches will end with a semicolon.

Save your sketch again to make sure that you don't lose any of your work.

The Loop Function

Remember that our goal is to make the LED blink repeatedly. To do this, we'll create a loop function to tell the Arduino to execute an instruction over and over until the power is shut off or someone presses the RESET button.

Enter the code shown in boldface after the void setup() section in the following listing to create an empty loop function. Be sure to end this new section with another brace (}), and then save your sketch again.

```
/*
Arduino Blink LED Sketch
by Mary Smith, created 07/01/13
*/

void setup()
{
  pinMode(13, OUTPUT); // set digital pin 13 to output
}
void loop()
{
  // place your main loop code here:
}
```

The Arduino IDE does not automatically save sketches, so save your work frequently!

Next, enter the actual functions into void loop() for the Arduino to execute.

Enter the following between the loop function's braces, and then click **Verify** to make sure that you've entered everything correctly:

```
digitalWrite(13, HIGH); // turn on digital pin 13
delay(1000); // pause for one second
digitalWrite(13, LOW); // turn off digital pin 13
delay(1000); // pause for one second
```

Let's take this all apart. The digitalWrite() function controls the voltage that is output from a digital pin: in this case, pin 13 to the LED. By setting the second parameter of this function to HIGH, a "high" digital voltage is output; then current will flow from the pin and the LED will turn on. (If you were to set this parameter to LOW, then the current flowing through the LED would stop.)

With the LED turned on, the light pauses for 1 second with delay(1000). The delay() function causes the sketch to do nothing for a period of time—in this case, 1,000 milliseconds, or 1 second.

Next, we turn off the voltage to the LED with digitalWrite(13, LOW);. Finally, we pause again for 1 second while the LED is off, with delay(1000);.

The completed sketch should look like this:

```
/*
 Arduino Blink LED Sketch
 by Mary Smith, created 07/01/13
*/

void setup()
{
  pinMode(13, OUTPUT); // set digital pin 13 to output
}
```

```
void loop()
{
  digitalWrite(13, HIGH); // turn on digital pin 13
  delay(1000); // pause for one second
  digitalWrite(13, LOW); // turn off digital pin 13
  delay(1000); // pause for one second
}
```

Before you do anything further, save your sketch!

Verifying Your Sketch

When you verify your sketch, you ensure that it has been written correctly in a way that the Arduino can understand. To verify your complete sketch, click **Verify** in the IDE and wait a moment. Once the sketch has been verified, a note should appear in the message window, as shown in Figure 2-12.

Figure 2-12: The sketch has been verified.

This "Done compiling" message tells you that the sketch is okay to upload to your Arduino. It also shows how much memory it will use (1,076 bytes in this case) of the total available on the Arduino (32,256 bytes).

But what if your sketch isn't okay? Say, for example, you forgot to add a semicolon at the end of the second delay(1000) function. If something is broken in your sketch, then when you click **Verify**, the message window should display a verification error message similar to the one shown in Figure 2-13.

Figure 2-13: The message window with a verification error

The message tells you that the error occurs in the void loop function, lists the line number of the sketch where the IDE thinks the error is located (blinky:16, or line 16 of your *blinky* sketch), and displays the error itself (the missing semicolon, error: expected ';' before '}' token). Furthermore, the IDE should also highlight in yellow the location of the error or a spot just after it. This helps you easily locate and rectify the mistake.

Uploading and Running Your Sketch

Once you're satisfied that your sketch has been entered correctly, save it, ensure that your Arduino board is connected, and click **Upload** in the IDE. The IDE may verify your sketch again and then upload it to your Arduino board. During this process, the TX/RX LEDs on your board (shown in Figure 2-6) should blink, indicating that information is traveling between the Arduino and your computer.

Now for the moment of truth: Your Arduino should start running the sketch. If you've done everything correctly, then the LED should blink on and off once every second!

Congratulations. You now know the basics of how to enter, verify, and upload an Arduino sketch.

Modifying Your Sketch

After running your sketch, you may want to change how it operates, by, for example, adjusting the on or off delay time for the LED. Because the IDE is a lot like a word processor, you can open your saved sketch, adjust the values, and then save your sketch again and upload it to the Arduino. For example, to increase the rate of blinking, change both delay functions to make the LEDs blink for one-quarter of a second by adjusting the delay to 250 like this:

```
delay(250); // pause for one-quarter of one second
```

Then upload the sketch again. The LED should now blink faster, for one-quarter of a second each time.

Looking Ahead

Armed with your newfound knowledge of how to enter, edit, save, and upload Arduino sketches, you're ready for the next chapter, where you'll learn how to use more functions, implement good project design, construct basic electronic circuits, and do much more.

3

FIRST STEPS

In this chapter you will

- Learn the concepts of good project design
- Learn the basic properties of electricity
- Be introduced to the resistor, light-emitting diode (LED), transistor, rectifier diode, and relay
- Use a solderless breadboard to construct circuits
- Learn how integer variables, for loops, and digital outputs can be used to create various LED effects

Now you'll begin to bring your Arduino to life. As you will see, there is more to working with Arduino than just the board itself. You'll learn how to plan projects in order to make your ideas a reality and then move on to a quick primer on electricity. Electricity is the driving force behind everything we do in this book, and it's important to have a solid understanding of the basics in order to create your own projects. You'll also take a look at the components that help bring real projects to life. Finally, you'll examine some new functions that are the building blocks for your Arduino sketches.

Planning Your Projects

When starting your first few projects, you might be tempted to write your sketch immediately after you've come up with a new idea. But before you start writing, a few basic preparatory steps are in order. After all, your Arduino board isn't a mind-reader; it needs precise instructions, and even if these instructions can be executed by the Arduino, the results may not be what you expected if you overlooked even a minor detail.

Whether you are creating a project that simply blinks a light or an automated model railway signal, a detailed plan is the foundation of success. When designing your Arduino projects, follow these basic steps:

1. **Define your objective.** Determine what you want to achieve.

2. **Write your algorithm.** An *algorithm* is a set of instructions that describes how to accomplish your project. Your algorithm will list the steps necessary for you to achieve your project's objective.

3. **Select your hardware.** Determine how it will connect to the Arduino.

4. **Write your sketch.** Create your initial program that tells the Arduino what to do.

5. **Wire it up.** Connect your hardware, circuitry, and other items to the Arduino board.

6. **Test and debug.** Does it work? During this stage, you identify errors and find their causes, whether in the sketch, hardware, or algorithm.

The more time you spend planning your project, the easier time you'll have during the testing and debugging stage.

NOTE *Even well-planned projects sometimes fall prey to feature creep. Feature creep occurs when people think up new functionality that they want to add to a project and then try to force new elements into an existing design. When you need to change a design, don't try to "slot in" or modify it with 11th-hour additions. Instead, start fresh by redefining your objective.*

About Electricity

Let's spend a bit of time discussing electricity, since you'll soon be building electronic circuits with your Arduino projects. In simple terms, *electricity* is a form of energy that we can harness and convert into heat, light, movement, and power. Electricity has three main properties that will be important to us as we build projects: current, voltage, and power.

Current

The flow of electrical energy through a circuit is called the *current*. Electrical current flows through a *circuit* from the positive side of a power source, such as a battery, to the negative side of the power source. This is known as *direct current (DC)*. For the purposes of this book, we will not deal with

AC (alternating current). In some circuits, the negative side is called *ground (GND)*. Current is measured in *amperes* or "amps" (*A*). Small amounts of current are measured in *milliamps (mA)*, where 1,000 milliamps equal 1 amp.

Voltage

Voltage is a measure of the difference in potential energy between a circuit's positive and negative ends. This is measured in *volts (V)*. The greater the voltage, the faster the current moves through a circuit.

Power

Power is a measurement of the rate at which an electrical device converts energy from one form to another. Power is measured in *watts (W)*. For example, a 100 W light bulb is much brighter than a 60 W bulb because the higher-wattage bulb converts more electrical energy into light.

A simple mathematical relationship exists among voltage, current, and power:

$$\text{Power (W)} = \text{Voltage (V)} \times \text{Current (A)}$$

Electronic Components

Now that you know a little bit about the basics of electricity, let's look at how it interacts with various electronic components and devices. Electronic *components* are the various parts that control electric current in a circuit to make our designs a reality. Just as the various parts of a car work together to provide fuel, power, and mobility to allow us to drive, electronic components work together to control and harness electricity to help us create useful devices.

Throughout this book, I'll explain specialized components as we use them. The following sections describe some of the fundamental components.

The Resistor

Various components, such as the Arduino's LED, require only a small amount of current to function—usually around 10 mA. When the LED receives excess current, it converts the excess to heat—too much of which can kill an LED. To reduce the flow of current to components such as LEDs, we can add a *resistor* between the voltage source and the component. Current flows freely along normal copper wire, but when it encounters a resistor, its movement is slowed. Some current is converted into a small amount of heat energy, which is proportional to the value of the resistor. Figure 3-1 shows an example of commonly used resistors.

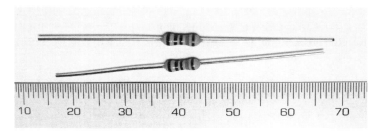

Figure 3-1: Typical resistors

Resistance

The level of resistance can be either fixed or variable. Resistance is measured in *ohms* (Ω) and can range from zero to thousands of ohms (*kiloohms*, or kΩ) to millions of ohms (*megohms*, or MΩ).

Reading Resistance Values

Resistors are very small, so their resistance value usually cannot be printed on the components themselves. Although you can test resistance with a multimeter, you can also read resistance directly from a physical resistor, even without numbers. One common way to show the component's resistance is with a series of color-coded bands, read from left to right, as follows:

First band   Represents the first digit of the resistance

Second band   Represents the second digit of the resistance

Third band   Represents the multiplier (for four-band resistors) or the third digit (for five-band resistors)

Fourth band   Represents the multiplier for five-band resistors

Fifth band   Shows the tolerance (accuracy)

Table 3-1 lists the colors of resistors and their corresponding values.

The fifth band represents a resistor's *tolerance*. This is a measure of the accuracy of the resistor. Because it is difficult to manufacture resistors with exact values, you select a margin of error as a percentage when buying a resistor. A brown band indicates 1 percent, gold indicates 5 percent, and silver indicates 10 percent tolerance.

Figure 3-2 shows a resistor diagram. The yellow, violet, and orange resistance bands are read as 4, 7, and 3, respectively, as listed in Table 3-1. These values translate to 47,000 Ω, more commonly read as 47 kΩ.

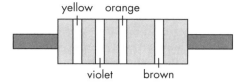

Figure 3-2: Example resistor diagram

Table 3-1: Values of Bands Printed on a Resistor, in Ohms

Color	Ohms
Black	0
Brown	1
Red	2
Orange	3
Yellow	4
Green	5
Blue	6
Violet	7
Gray	8
White	9

Chip Resistors

Surface-mount chip resistors display a printed number and letter code, as shown in Figure 3-3, instead of color stripes. The first two digits represent a single number, and the third digit represents the number of zeros to follow that number. For example, the resistor in Figure 3-3 has a value of 10,000 Ω, or 10 kΩ.

Figure 3-3: Example of a surface-mount resistor

NOTE   *If you see a number and letter code on small chip resistors (such as* 01C*), google* EIA-96 code calculator *for lookup tables on that more involved code system.*

Multimeters

A *multimeter* is an incredibly useful and relatively inexpensive piece of test equipment that can measure voltage, resistance, current, and more. For example, Figure 3-4 shows a multimeter measuring a resistor.

Figure 3-4: Multimeter measuring a 560-ohm 1 percent tolerance resistor

If you are colorblind, a multimeter is essential. As with other good tools, purchase your multimeter from a reputable retailer instead of fishing about on the Internet for the cheapest one you can find.

Power Rating

The resistor's *power rating* is a measurement of the power, in watts, that it will tolerate before overheating or failing. The resistors shown in Figure 3-1 are 1/4-watt resistors, which are the most commonly used resistors with the Arduino system.

When you're selecting a resistor, consider the relationship among power, current, and voltage. The greater the current and/or voltage, the greater the resistor's power.

Usually, the greater a resistor's power rating, the greater its physical size. For example, the resistor shown in Figure 3-5 is a 5-watt resistor, which measures 26 mm long by 7.5 mm wide.

Figure 3-5: A 5-watt resistor

The Light-Emitting Diode

The LED is a very common, infinitely useful component that converts electrical current into light. LEDs come in various shapes, sizes, and colors. Figure 3-6 shows a common LED.

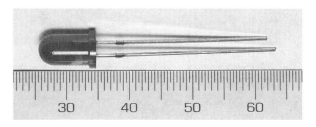

Figure 3-6: Red LED, 5 mm in diameter

Connecting LEDs in a circuit takes some care, because they are *polarized*; this means that current can enter and leave the LED in one direction only. The current enters via the *anode* (positive) side and leaves via the *cathode* (negative) side, as shown in Figure 3-7. Any attempt to make too much current flow through an LED in the opposite direction will break the component.

Thankfully, LEDs are designed so that you can tell which end is which. The leg on the anode side is longer, and the rim at the base of the LED is flat on the cathode side, as shown in Figure 3-8.

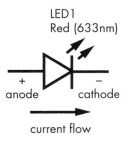

Figure 3-7: Current flow through an LED

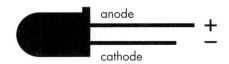

Figure 3-8: LED design indicates the anode (longer leg) and cathode (flat rim) sides.

When adding LEDs to a project, you need to consider the operating voltage and current. For example, common red LEDs require around 1.7 V and 5 to 20 mA of current. This presents a slight problem for us, because the Arduino outputs a set 5 V and a much higher current. Luckily, we can use a *current-limiting resistor* to reduce the current flow into an LED. But which value resistor do we use? That's where Ohm's Law comes in.

To calculate the required current-limiting resistor for an LED, use this formula:

$$R = (V_s - V_f) \div I$$

where V_s is the supply voltage (Arduino outputs 5 V); V_f is the LED forward voltage drop (say, 1.7 V), and I is the current required for the LED (10 mA). (The value of I must be in amps, so 10 mA converts to 0.01 A.)

Now let's use this for our LEDs—with a value of 5 V for V_s, 1.7 V for V_f, and 0.01 A for I. Substituting these values into the formula gives a value for R of 330 Ω. However, the LEDs will happily light up when fed current less than 10 mA. It's good practice to use lower currents when possible to protect sensitive electronics, so we'll use 560 Ω, 1/4-watt resistors with our LEDs, which allow around 6 mA of current to flow.

NOTE *When in doubt, always choose a slightly higher value resistor, because it's better to have a dim LED than a dead one!*

THE OHM'S LAW TRIANGLE

Ohm's Law states that the relationship between current, resistance, and voltage is as follows:

voltage (V) = current (I) × resistance (R)

If you know two of the quantities, then you can calculate the third. A popular way to remember Ohm's Law is with a triangle, as shown in Figure 3-9.

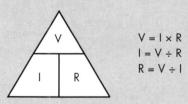

$$V = I \times R$$
$$I = V \div R$$
$$R = V \div I$$

Figure 3-9: The Ohm's Law triangle

The Ohm's Law triangle diagram is a convenient tool for calculating voltage, current, or resistance when two of the three values are known. For example, if you need to calculate resistance, put your finger over *R*, which leaves you with voltage divided by current; to calculate voltage, cover *V*, which leaves you with current times resistance.

The Solderless Breadboard

Our ever-changing circuits will need a base—something to hold them together and build upon. A great tool for this purpose is a *solderless breadboard*. The breadboard is a plastic base with rows of electrically connected sockets (just don't cut bread on them). They come in many sizes, shapes, and colors, as shown in Figure 3-10.

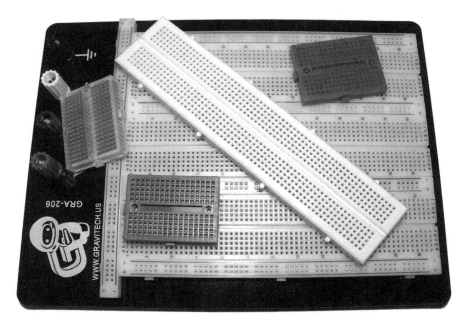

Figure 3-10: Breadboards in various shapes and sizes

The key to using a breadboard is knowing how the sockets are connected—whether in short columns or in long rows along the edge or in the center. The connections vary by board. For example, in the breadboard shown at the top of Figure 3-11, columns of five holes are connected vertically but isolated horizontally. If you place two wires in one vertical row, then they will be electrically connected. By the same token, the long rows in the center between the horizontal lines are connected horizontally. Because you'll often need to connect a circuit to the supply voltage and ground, these long horizontal lines of holes are ideal for the supply voltage and ground.

When you're building more complex circuits, a breadboard will get crowded and you won't always be able to place components exactly where you want. It's easy to solve this problem using short connecting wires, however. Retailers that sell breadboards usually also sell small boxes of wires of various lengths, such as the assortment shown in Figure 3-12.

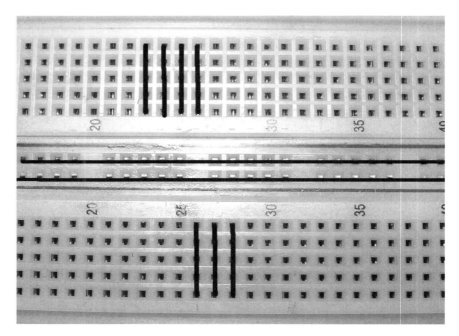

Figure 3-11: Breadboard internal connections

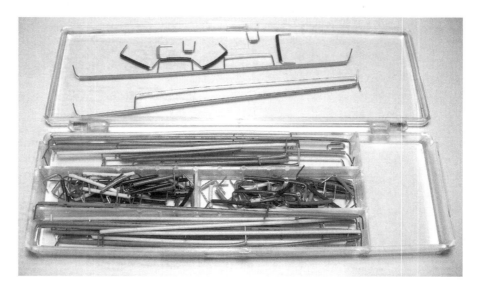

Figure 3-12: Assorted breadboard wires

Project #1: Creating a Blinking LED Wave

Let's put some LEDs and resistors to work. In this project, we'll use five LEDs to emulate the front of the famous TV show vehicle KITT from the television show *Knight Rider*, creating a kind of wavelike light pattern.

The Algorithm

Here's our algorithm for this project:

1. Turn on LED 1.
2. Wait half a second.
3. Turn off LED 1.
4. Turn on LED 2.
5. Wait half a second.
6. Turn off LED 2.
7. Continue until LED 5 is turned on, at which point the process reverses from LED 5 to 1.
8. Repeat indefinitely.

The Hardware

Here's what you'll need to create this project:

- Five LEDs
- Five 560 Ω resistors
- One breadboard
- Various connecting wires
- Arduino and USB cable

We will connect the LEDs to digital pins 2 through 6 via the 560-ohm current-limiting resistors.

The Sketch

Now for our sketch. Enter this code into the IDE:

```
// Project 1 - Creating a Blinking LED Wave
void setup()
{
  pinMode(2, OUTPUT);    // LED 1 control pin is set up as an output
  pinMode(3, OUTPUT);    // same for LED 2 to LED 5
  pinMode(4, OUTPUT);
  pinMode(5, OUTPUT);
  pinMode(6, OUTPUT);
}
```

```
❷ void loop()
{
  digitalWrite(2, HIGH);  // Turn LED 1 on
  delay(500);             // wait half a second
  digitalWrite(2, LOW);   // Turn LED 1 off
  digitalWrite(3, HIGH);  // and repeat for LED 2 to 5
  delay(500);
  digitalWrite(3, LOW);
  digitalWrite(4, HIGH);
  delay(500);
  digitalWrite(4, LOW);
  digitalWrite(5, HIGH);
  delay(500);
  digitalWrite(5, LOW);
  digitalWrite(6, HIGH);
  delay(500);
  digitalWrite(6, LOW);
  digitalWrite(5, HIGH);
  delay(500);
  digitalWrite(5, LOW);
  digitalWrite(4, HIGH);
  delay(500);
  digitalWrite(4, LOW);
  digitalWrite(3, HIGH);
  delay(500);
  digitalWrite(3, LOW);
  // the loop() will now loop around and start from the top again
}
```

In void setup() at ❶, the digital I/O pins are set to outputs, because we want them to send current to the LEDs on demand. We specify when to turn on each LED using the digitalWrite() function in the void loop() ❷ section of the sketch.

The Schematic

Now let's build the circuit. Circuit layout can be described in several ways. For the first few projects in this book, we'll use physical layout diagrams similar to the one shown in Figure 3-13.

By comparing the wiring diagram to the functions in the sketch, you can begin to make sense of the circuit. For example, when we use digitalWrite(2, HIGH), a high voltage of 5 V flows from digital pin 2, through the current-limiting resistor, through the LED via the anode and then the cathode, and finally back to the Arduino's GND socket to complete the circuit. Then, when we use digitalWrite(2, LOW), the current stops and the LED turns off.

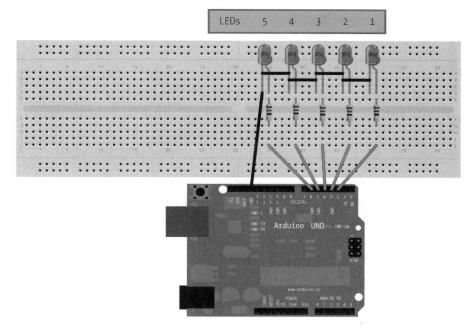

Figure 3-13: Circuit layout for Project 1

Running the Sketch

Now connect your Arduino and upload the sketch. After a second or two, the LEDs should blink from left to right and then back again. Success is a wonderful thing—embrace it!

If nothing happens, however, then immediately remove the USB cable from the Arduino and check that you typed the sketch correctly. If you find an error, fix it and upload your sketch again. If your sketch matches exactly and the LEDs still don't blink, check your wiring on the breadboard.

You now know how to make an LED blink with your Arduino, but this sketch is somewhat inefficient. For example, if you wanted to modify this sketch to make the LEDs cycle more quickly, you would need to alter each delay(500). There is a better way.

Using Variables

In computer programs, we use *variables* to store data. For example, in the sketch for Project 1, we used the function delay(500) to keep the LEDs turned on.

The problem with the sketch as written is that it's not very flexible. If we want to make a change to the delay time, then we have to change each entry manually. To address this problem, we'll create a variable to represent the value for the delay() function.

Enter the following line in the Project 1 sketch above the void setup() function and just after the initial comment:

```
int d = 250;
```

This assigns the number 250 to a variable called d.

Next, change every 500 in the sketch to a d. Now when the sketch runs, the Arduino will use the value in d for the delay() functions. When you upload the sketch after making these changes, the LEDs will turn on and off at a much faster rate, as the delay value is much smaller at the 250 value.

int indicates that the variable contains an integer—a whole number between −32,768 and 32,767. Simply put, any integer value has no fraction or decimal places. Now, to alter the delay, simply change the variable declaration at the start of the sketch. For example, entering 100 for the delay would speed things up even more:

```
int d = 100;
```

Experiment with the sketch, perhaps altering the delays and the sequence of HIGH and LOW. Have some fun with it. Don't disassemble the circuit yet, though; we'll continue to use it with more projects in this chapter.

Project #2: Repeating with for Loops

When designing a sketch, you'll often repeat the same function. You could simply copy and paste the function to duplicate it in a sketch, but that's inefficient and a waste of your Arduino's program memory. Instead, you can use for loops. The benefit of using a for loop is that you can determine how many times the code inside the loop will repeat.

To see how a for loop works, enter the following code as a new sketch:

```
// Project 2 - Repeating with for Loops
int d = 100;

void setup()
{
  pinMode(2, OUTPUT);
  pinMode(3, OUTPUT);
  pinMode(4, OUTPUT);
  pinMode(5, OUTPUT);
  pinMode(6, OUTPUT);
}

void loop()
{
```

```
    for ( int a = 2; a < 7 ; a++ )
    {
      digitalWrite(a, HIGH);
      delay(d);
      digitalWrite(a, LOW);
      delay(d);
    }
}
```

The for loop will repeat the code within the curly brackets beneath it as long as some condition is true. Here, we have used a new integer variable, a, which starts with the value 2. Every time the code is executed, the a++ will add 1 to the value of a. The loop will continue in this fashion while the value of a is less than 7 (the condition). Once it is equal to or greater than 7, the Arduino moves on and continues with whatever code comes after the for loop.

The number of loops that a for loop executes can also be set by counting down from a higher number to a lower number. To demonstrate this, add the following loop to the Project 2 sketch after the first for loop:

```
❶ for ( int a = 5 ; a > 1 ; a-- )
    {
      digitalWrite(a, HIGH);
      delay(d);
      digitalWrite(a, LOW);
      delay(d);
    }
```

Here, the for loop at ❶ sets the value of a equal to 5 and then subtracts 1 after every loop due to the a--. The loop continues in this manner while the value for a is greater than 1 (a > 1) and finishes once the value of a falls to 1 or less than 1.

We have now re-created Project 1 using less code. Upload the sketch and see for yourself!

Varying LED Brightness with Pulse-Width Modulation

Rather than just turning LEDs on and off rapidly using digitalWrite(), we can define the level of brightness of an LED by adjusting the amount of time between each LED's on and off states using *pulse-width modulation (PWM)*. PWM can be used to create the illusion of an LED being on at different levels of brightness by turning the LED on and off rapidly, at around 500 cycles per second. The brightness we perceive is determined by the amount of time the digital output pin is on versus the amount of time it is off—that is, every time the LED is lit or unlit. Because our eyes can't see flickers faster than 50 cycles per second, the LED appears to have a constant brightness.

The greater the *duty cycle* (the longer the pin is on compared to off in each cycle), the greater the perceived brightness of the LED connected to the digital output pin.

Figure 3-14 shows various PWM duty cycles. The filled-in gray areas represent the amount of time that the light is on. As you can see, the amount of time per cycle that the light is on increases with the duty cycle.

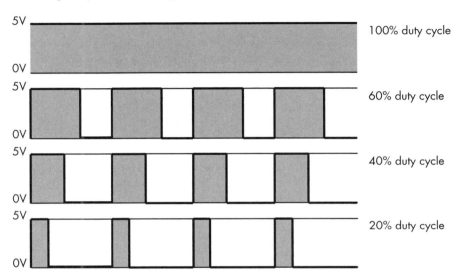

Figure 3-14: Various PWM duty cycles

Only digital pins 3, 5, 6, 9, 10, and 11 on a regular Arduino board can be used for PWM. They are marked on the Arduino board with a tilde (~), as shown in Figure 3-15.

Figure 3-15: The PWM pins are marked with a tilde (~).

To create a PWM signal, we use the function analogWrite(x, y), where x is the digital pin and y is a value for the duty cycle, between 0 and 255, where 0 indicates a 0 percent duty cycle and 255 indicates 100 percent duty cycle.

Project #3: Demonstrating PWM

Now let's try this with our circuit from Project 2. Enter the following sketch into the IDE and upload it to the Arduino:

```
// Project 3 - Demonstrating PWM
int d = 5;
void setup()
{
  pinMode(3, OUTPUT);    // LED control pin is 3, a PWM capable pin
}

void loop()
{
  for ( int a = 0 ; a < 256 ; a++ )
  {
    analogWrite(3, a);
    delay(d);
  }

  for ( int a = 255 ; a >= 0 ; a-- )
  {
    analogWrite(3, a);
    delay(d);
  }
  delay(200);
}
```

The LED on digital pin 3 will exhibit a "breathing effect" as the duty cycle increases and decreases. In other words, the LED will turn on, increasing in brightness until fully lit, and then reverse. Experiment with the sketch and circuit. For example, make five LEDs breathe at once, or have them do so sequentially.

More Electric Components

You'll usually find it easy to plan on having a digital output control do something without taking into account how much current the control really needs to get the job done. As you create your project, remember that each digital output pin on the Arduino Uno can offer a maximum of 40 mA of current per pin and 200 mA total for all pins. Three electronic hardware components can help you increase the current-handling ability of the Arduino, however, and are discussed next.

WARNING   *If you attempt to exceed 40 mA on a single pin, or 200 mA total, then you risk permanently damaging the microcontroller integrated circuit (IC).*

The Transistor

Almost everyone has heard of a *transistor*, but most people don't really understand how it works. In the spirit of brevity, I will keep the explanation as simple as possible. A transistor can turn on or off the flow of a much larger current than the Arduino Uno can handle. We can, however, safely control a transistor using an Arduino digital output pin. A popular example is the BC548, shown in Figure 3-16.

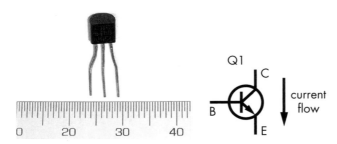

Figure 3-16: A typical transistor: the BC548

Similar to the LED, the transistor's pins have a unique function and need to be connected in the proper orientation. With the flat front of the transistor facing you (as shown on the left of Figure 3-16), the pins on the BC548 are called (from left to right) *collector*, *base*, and *emitter*. (Note that this pin order, or pinout, is for the BC548 transistor; other transistors may be oriented differently.) When a small current is applied to the base, such as from an Arduino digital I/O pin, the larger current we want to switch enters through the collector; then it is combined with the small current from the base, and then it flows out via the emitter. When the small control current at the base is turned off, no current can flow through the transistor.

The BC548 can switch up to 100 mA of current at a maximum of 30 V—much more than the Arduino's digital output. In projects later in the book, we will use this in other transistors, and at that time, you'll read about transistors in more detail.

NOTE   *Always pay attention to the pin order for your particular transistor, because each transistor can have its own orientation.*

The Rectifier Diode

The *diode* is a very simple yet useful component that allows current to flow in one direction only. It looks a lot like a resistor, as you can see in Figure 3-17.

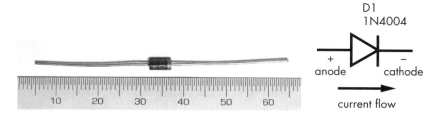

Figure 3-17: A 1N4004-type rectifier diode

The projects in this book will use the 1N4004-type rectifier diode. Current flows into the diode via the anode and out through the cathode, which is marked with the ring around the diode's body. These diodes can protect parts of the circuit against reverse current flow, but there is a price to pay: diodes also cause a drop in the voltage of around 0.7 V. The 1N4004 diode is rated to handle 1 A and 400 V, much higher than we will be using. It's a tough, common, and low-cost diode.

The Relay

Relays are used for the same reason as transistors—to control a much larger current and voltage. A relay has the advantage of being *electrically isolated* from the control circuit, which allows the Arduino to switch very large currents and voltages. Isolation is sometimes necessary to protect circuits from these very large currents and voltages, which can damage an Arduino. Inside the relay is an interesting pair of items: mechanical switch contacts and a low-voltage coil of wire, as shown in Figure 3-18.

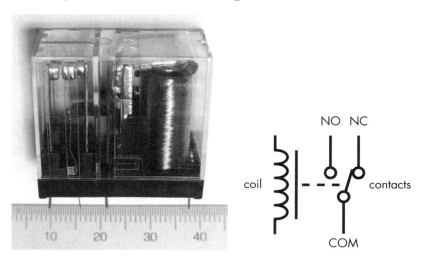

Figure 3-18: Inside a typical relay

When a current is applied to the coil, the component becomes an electromagnet and attracts a bar of metal that acts just like the toggle of a switch. The magnet pulls the bar in one direction when on and lets it fall back when off, thereby turning it on or off as current is applied and removed from the coil. This movement has a distinctive "click" that you might recognize from the turn signal in older cars.

Higher-Voltage Circuits

Now that you understand a bit about the transistor, rectifier diode, and relay, let's use them together to control higher currents and voltages. Connecting the components is very simple, as shown in Figure 3-19.

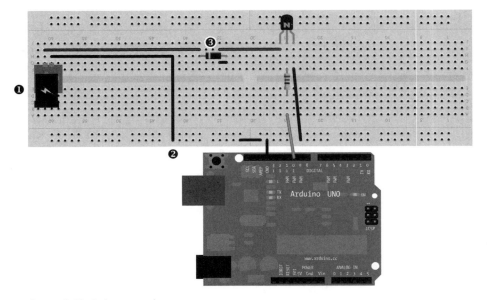

Figure 3-19: Relay control circuit

This simple example circuit controls a relay that has a 12 V coil. One use for this circuit would be to control a lamp or cooling fan connected to the relay switching contacts. The Arduino's digital pin 10 is connected to the transistor's base via a 1 kΩ resistor. The transistor controls the current through the relay's coil by switching it on and off. Remember that the pins are C, B, and then E when looking at the flat surface of the transistor. The object on the left of the breadboard at ❶ represents a 12 V power supply for the relay coil. The negative or ground at ❷ from the 12 V supply, the transistor's emitter pin, and Arduino GND are all connected together. Finally, a 1N4004 rectifier diode is connected across the relay's coil at ❸, with the cathode on the positive supply side. You can check the relay's data sheet to determine the pins for the contacts and to connect the controlled item appropriately.

The diode is in place to protect the circuit. When the relay coil changes from on to off, stray current remains briefly in the coil and becomes a high-voltage spike that has to go somewhere. The diode allows the stray current to loop around through the coil until it is dissipated as a tiny amount of heat. It prevents the turn-off spike from damaging the transistor or Arduino pin.

WARNING   *If you want to control mains-rated electricity (110–250 V) at a high current with a relay, contact a licensed electrician to complete this work for you. Even the slightest mistake can be fatal.*

Looking Ahead

And now Chapter 3 draws to a close. I hope you had fun trying out the examples and experimented with LED effects. In this chapter, you got to create blinking LEDs on the Arduino in various ways, did a bit of hacking, and learned how functions and loops can be used efficiently to control components connected to the Arduino. Studying this chapter has set you up for more success in the forthcoming chapters.

Chapter 4 will be a lot of fun. You will create some actual projects, including traffic lights, a thermometer, a battery tester, and more—so when you're ready to take it to the next level, turn the page!

4

BUILDING BLOCKS

In this chapter you will

- Learn how to read schematic diagrams, the language of electronic circuits
- Be introduced to the capacitor
- Work with input pins
- Use arithmetic and test values
- Make decisions with if-then-else statements
- Learn the difference between analog and digital
- Measure analog voltage sources at different levels of precision
- Be introduced to variable resistors, piezoelectric buzzers, and temperature sensors
- Consolidate your knowledge by creating traffic lights, a battery tester, and a thermometer

The information in this chapter will help you understand an Arduino's potential. We'll continue to learn more about electronics, including information about new components, how to read schematic diagrams (the "road maps" of electronic circuits), and the types of signals that can be measured. Then, we'll discuss additional Arduino functions—such as storing values, performing mathematical operations, and making decisions. Finally, we'll examine some more components and then put them to use in some useful projects.

Using Schematic Diagrams

Chapter 3 described how to build a circuit using physical layout diagrams to represent the breadboard and components mounted on it. Although such physical layout diagrams may seem like the easiest way to diagram a circuit, you'll find that as more components are added, direct representations can make physical diagrams a real mess. Because our circuits are about to get more complicated, we'll start using *schematic diagrams* (also known as *circuit diagrams*) to illustrate them, such as the one shown in Figure 4-1.

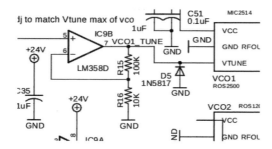

Figure 4-1: Example of a schematic diagram

Schematics are simply circuit "road maps" that show the path of electrical current flowing through various components. Instead of showing components and wires, a schematic uses symbols and lines.

Identifying Components

Once you know what the symbols mean, reading a schematic is easy. To begin, let's examine the symbols for the components we've already used.

The Arduino

Figure 4-2 shows a symbol for the Arduino itself. As you can see, all of the Arduino's connections are displayed and neatly labeled.

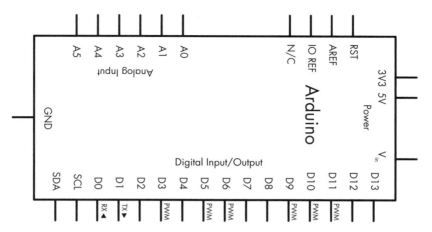

Figure 4-2: Arduino Uno symbol

The Resistor

The resistor symbol is shown in Figure 4-3.

It's good practice to display the resistor value and part designator along with the resistor symbol (220 Ω and R1 in this case). This makes life a lot easier for everyone trying to make sense of the schematic (including you). Often you may see ohms written as *R* instead—for example, 220 R.

Figure 4-3: Resistor symbol

The Rectifier Diode

The rectifier diode is shown in Figure 4-4.

Recall from Chapter 3 that rectifier diodes are polarized, and current flows from the anode to the cathode. On the symbol shown in Figure 4-4, the anode is on the left and the cathode is on the right. An easy way to remember this is to think of current flowing toward the point of the triangle only. Current cannot flow the other way, because the vertical bar "stops" it.

D1
1N4004

+ anode — cathode

current flow

Figure 4-4: Rectifier diode symbol

The LED

The LED symbol is shown in Figure 4-5.

All members of the diode family share a common symbol: the triangle and vertical line. However, LED symbols show two parallel arrows pointing away from the triangle to indicate that light is being emitted.

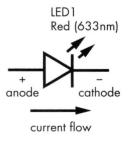

Figure 4-5: LED symbol

The Transistor

The transistor symbol is shown in Figure 4-6. We'll use this to represent our BC548.

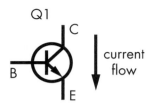

Q1

current flow

The vertical line at the top of the symbol (labeled *C*) represents the collector, the horizontal line at the left represents the base (labeled *B*), and the bottom line represents the emitter (labeled *E*). The arrow inside the symbol, pointing down and to the right, tells us that this is an *NPN*-type transistor, because NPN transistors allow current to flow from the collector to the emitter. (*PNP*-type transistors allow current to flow from the emitter to collector.)

When numbering transistors, we use the letter *Q*, just as we use *R* to number resistors.

Figure 4-6: Transistor symbol

The Relay

The relay symbol is shown in Figure 4-7.

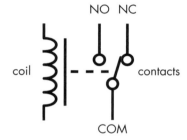

NO NC

coil   contacts

COM

Relay symbols can vary in many ways and may have more than one set of contacts, but all relay symbols share certain elements in common. The first is the *coil*, which is the curvy vertical line at the left. The second element is the relay *contacts*. The *COM* (for common) contact is often used as an input, and the contacts marked *NO* (normally open) and *NC* (normally closed) are often used as outputs.

The relay symbol is always shown with the relay in the off state and the coil not *energized*—that is, with the COM and NC pins connected. When the relay coil is energized, the COM and NO pins will be connected in the symbol.

Figure 4-7: Relay symbol

Wires in Schematics

When wires cross or connect in schematics, they are drawn in particular ways, as shown in the following examples.

Crossing but Not Connected Wires

When two wires cross but are not connected, the crossing can be represented in one of two ways, as shown in Figure 4-8. There is no one right way; it's a matter of preference.

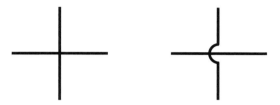

Figure 4-8: Nonconnecting crossed wires

Connected Wires

When wires are meant to be physically connected, a *junction dot* is drawn at the point of connection, as shown in Figure 4-9.

Figure 4-9: Two wires that are connected

Wire Connected to Ground

When a wire is connected back to ground (GND), the standard method is to use the symbol shown in Figure 4-10.

The GND symbol at the end of a line in a schematic tells you that the wire is physically connected to the Arduino GND pin.

Figure 4-10: The GND symbol

Dissecting a Schematic

Now that you know the symbols for various components and their connections, let's dissect the schematic we would draw for Project 1. Recall that you made five LEDs blink backward and forward.

Compare the schematic shown in Figure 4-11 with Figure 3-13 on page 45, and you'll probably agree that using a schematic is a much easier way to describe a circuit.

From now on, we'll use schematics to describe circuits, and we'll show the symbols for new components as they're introduced.

NOTE   *If you'd like to create your own computer-drawn schematics, try the Fritzing application, available for free from* http://www.fritzing.org/.

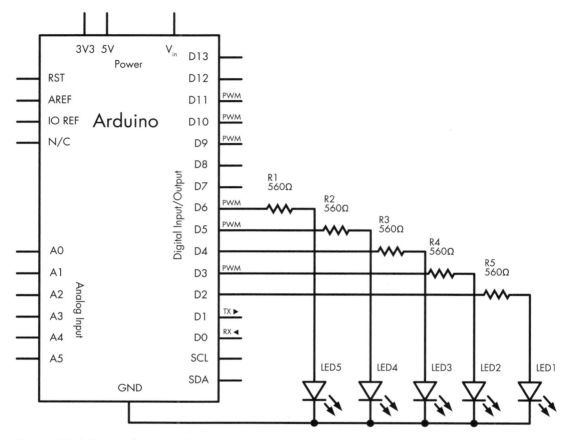

Figure 4-11: Schematic for Project 1

The Capacitor

A *capacitor* is a device that holds an electric charge. It consists of two metal plates with an insulating layer that allows an electric charge to build up between the plates. Once the current is stopped, the charge remains and can flow out of the capacitor (called *discharging* the capacitor) as soon as the charge voltage stored in the capacitor is presented with a new path for the current to take.

Measuring the Capacity of a Capacitor

The amount of charge that a capacitor can store is measured in *farads*, and one farad is actually a very large amount. Therefore, you will generally find capacitors with values measured in picofarads or microfarads.

One *picofarad* (pF) is 0.000000000001 of a farad, and one *microfarad* (µF) is 0.000001 of a farad. Capacitors are also manufactured to accept certain voltage maximums. In this book, we'll be working with low voltages only, so we won't be using capacitors rated at greater than 10 V or so; it's generally fine, however, to use higher-voltage specification capacitors in lower-voltage circuits. Common voltage ratings are 10, 16, 25, and 50 V.

Reading Capacitor Values

Reading the value of a ceramic capacitor takes some practice, because the value is printed in a sort of code. The first two digits represent the value in picofarads, and the third digit is the multiplier in tens. For example, the capacitor shown in Figure 4-12 is labeled *104*. This equates to 10, followed by four zeros, which equals 100,000 picofarads/pF (100 nanofarads [nF], or 0.1 microfarads [µF]).

Figure 4-12: A 0.1 µF ceramic capacitor

NOTE   *The conversions between units of measure can be a little confusing, but you can print an excellent conversion chart from* http://www .justradios.com/uFnFpF.html.

Types of Capacitors

Our projects will use two types of capacitors: ceramic and electrolytic.

Ceramic Capacitors

Ceramic capacitors, such as the one shown in Figure 4-12, are very small and therefore hold a small amount of charge. They are not polarized and can be used for current flowing in either direction. The schematic symbol for a nonpolarized capacitor is shown in Figure 4-13.

Ceramic capacitors work beautifully in high-frequency circuits because they can charge and discharge very quickly due to their small capacitance.

Figure 4-13: A nonpolarized capacitor schematic symbol, with the capacitor's value shown at the upper right

Electrolytic Capacitors

Electrolytic capacitors, like the one shown in Figure 4-14, are physically larger than ceramic types, offer increased capacitance, and are polarized. A marking on the cover shows either the positive (+) side or negative side (–). In Figure 4-14, you can see the stripe and the small negative (–) symbol that identifies the negative side. Like resistors, capacitors also have a level of tolerance with their values. The capacitor in Figure 4-14 has a tolerance of 20 percent and a capacitance of 100 µF.

Figure 4-14: Electrolytic capacitor

The schematic symbol for electrolytic capacitors, shown in Figure 4-15, includes the + symbol to indicate the capacitor's polarity.

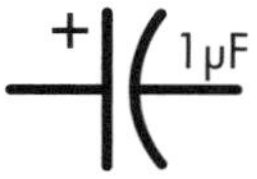

$$+\!\!-\!\!\!(\!\!- \quad 1\,\mu F$$

Figure 4-15: Polarized capacitor schematic symbol

Electrolytic capacitors are often used to store larger electric charges and to smooth power supply voltages. Like a small temporary battery, they can provide power-supply smoothing and stability near circuits or parts that draw high currents quickly from the supply. This prevents unwanted dropouts and noise in your circuits. Luckily, the values of the electrolytic capacitor are printed clearly on the outside and don't require decoding or interpretation.

Now that you have experience generating basic forms of output using LEDs with your Arduino, it's time to learn how to send input from the outside world into your Arduino using digital inputs and to make decisions based on that input.

Digital Inputs

In Chapter 3, we used digital I/O pins as outputs to turn LEDs on and off. We can use these same pins to accept input from users—such as detecting whether a push button has been pressed by a user.

Like digital outputs, digital inputs have two states: high and low. The simplest form of digital input is a push button, like those shown in Figure 4-16. You can insert these directly into your solderless breadboard. A *push button* allows a voltage or current to pass when the button is pressed, and digital input pins are used to detect the presence of the voltage and to determine whether a button is pressed.

Figure 4-16: Basic push buttons on a breadboard

Notice how the button at the bottom of the figure is inserted into the breadboard, bridging rows 23 and 25. When the button is pressed, it connects the two rows. The schematic symbol for this particular push button is shown in Figure 4-17. The symbol represents the two sides of the button, which are numbered with the prefix S. When the button is pressed, the line bridges the two halves and allows voltage or current through.

Figure 4-17: Push-button schematic symbol

MEASURING SWITCH BOUNCE
WITH A DIGITAL STORAGE OSCILLOSCOPE

Push buttons exhibit a phenomenon called *switch bounce*, or *bouncing*, which refers to a button's tendency to turn on and off several times after being pressed only once by the user. This phenomenon occurs because the metal contacts inside a push button are so small that they can vibrate after a button has been released, thereby switching on and off again very quickly.

Switch bounce can be demonstrated with a *digital storage oscilloscope (DSO)*, a device that displays the change in a voltage over a period of time. For example, consider Figure 4-18, a DSO displaying a switch bounce.

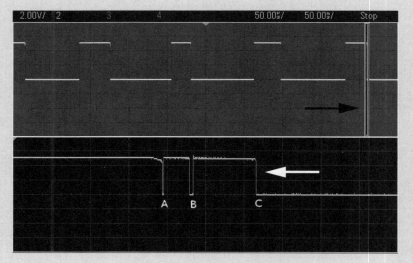

Figure 4-18: Measuring switch bounce

The top half of the display in Figure 4-18 shows the results of pressing a button several times. When the voltage line indicated by the arrows is at the higher horizontal position (5 V), the button is in the *on* state and the voltage is connected through it. Underneath the word *Stop* is a slice of time just after the button was switched off, as shown by two vertical lines. The button voltage during this time is magnified in the bottom half of the screen. At A, the button is released by the user and the line drops down to 0 V. However, as you can see, due to physical vibration, the button is again at the higher 5 V position until B, where it vibrates off and then on again until C, where it settles at the low (off) state. In effect, instead of relaying one button press to our Arduino, in this case, we have unwittingly sent three.

Project #4: Demonstrating a Digital Input

Our goal in this project is to create a button that turns on an LED for half a second when pressed.

The Algorithm

Here is our algorithm:

1. Test to see if the button has been pressed.
2. If the button has been pressed, then turn on the LED for half a second, and then turn it off.
3. If the button has not been pressed, then do nothing.
4. Repeat indefinitely.

The Hardware

Here's what you'll need to create this project:

- One push button
- One LED
- One 560 Ω resistor
- One 10 kΩ resistor
- One 100 nF capacitor
- Various connecting wires
- One breadboard
- Arduino and USB cable

The Schematic

First we create the circuit on the breadboard with the schematic shown in Figure 4-19. Notice how the 10 kΩ resistor is connected between GND and digital pin seven. We call this a *pull-down resistor*, because it pulls the voltage at the digital pin almost to zero. Furthermore, by adding a 100 nF capacitor across the 10 kΩ resistor, we create a simple *debounce* circuit to help filter out the switch bounce. When the button is pressed, the digital pin goes immediately to high. But when the button is released, digital pin seven is pulled down to GND via the 10 kΩ resistor, and the 100 nF capacitor creates a small delay. This effectively covers up the bouncing pulses by slowing down the voltage falling to GND, thereby eliminating most of the false readings due to floating voltage and erratic button behavior.

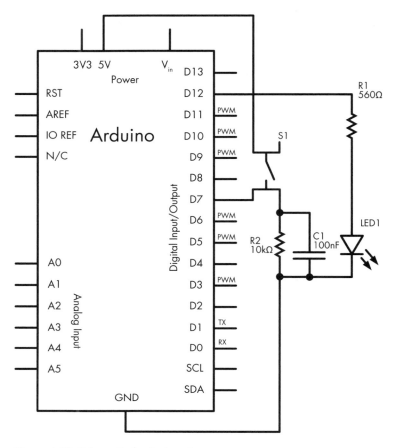

Figure 4-19: Schematic for Project 4

Because this is the first time you're building a circuit with a schematic, follow these step-by-step instructions as you walk through the schematic; this should help you understand how the components connect:

1. Insert the push button into the breadboard, as shown in Figure 4-20.

2. Turn the breadboard 90 degrees counterclockwise and insert the 10 kΩ resistor, a short link wire, and the capacitor, as shown in Figure 4-21.

Figure 4-20: Push button inserted into breadboard

Figure 4-21: 10 kΩ resistor, capacitor, and push button

3. Connect one wire from the Arduino 5 V pin to the leftmost vertical column on the breadboard. Connect another wire from the Arduino GND pin to the vertical row to the right of the 5 V column. Connect another horizontal wire between the vertical GND column and the bottom-left pin of the button. This is shown in Figure 4-22.

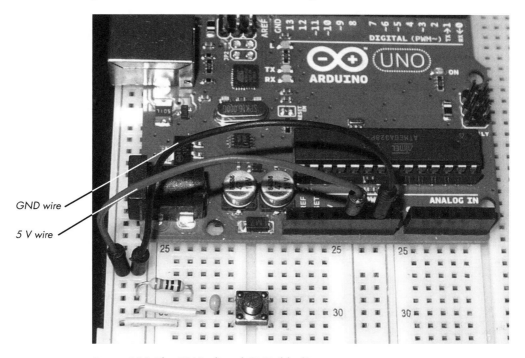

Figure 4-22: The 5 V (red) and GND (black) wires

4. Run a wire from Arduino digital pin 7 to the breadboard near the top-right corner of the button, as shown in Figure 4-23.

Figure 4-23: Connecting the button to the digital input

5. Insert the LED into the breadboard with the short leg (the cathode) connected to the GND column, and the long leg (the anode) in a row to the right. Next, connect the 560 Ω resistor to the right of the LED, as shown in Figure 4-24.

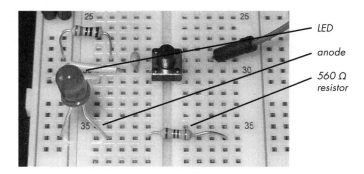

Figure 4-24: Inserting the LED and 560 Ω resistor

6. Connect a wire from the right side of the 560 Ω resistor to Arduino digital pin 12, as shown in Figure 4-25.

Figure 4-25: Connecting the LED branch to the Arduino

Before continuing, review the schematic for this circuit and check that your components are wired correctly. Compare the schematic against the actual wiring of the circuit.

The Sketch

For the sketch, enter and upload Listing 4-1:

```
// Project 4 - Demonstrating a Digital Input
❶ #define LED    12
   #define BUTTON 7

   void setup()
   {
❷    pinMode(LED, OUTPUT);    // output for the LED
      pinMode(BUTTON, INPUT); // input for the button
   }

   void loop()
   {
      if ( digitalRead(BUTTON) == HIGH )
      {
```

```
        digitalWrite(LED, HIGH);   // turn on the LED
        delay(500);                // wait for 0.5 seconds
        digitalWrite(LED, LOW);    // turn off the LED
    }
}
```

Listing 4-1: Digital input

After you've uploaded your sketch, tap the push button briefly and your
LED should stay on for half a second.

Modifying Your Sketch

Once you've had some success, try modifying your sketch by changing the
length of time that the light stays on or by adding a push button control to
Project 3. (Don't disassemble this circuit, though; we'll use it again in the
next example.)

Understanding the Sketch

Let's examine the new items in the sketch for Project 4—specifically, #define,
digital input pins, and the if-then function.

Creating Constants with #define

Before void setup(), we use #define statements at ❶ to create fixed variables:
When the sketch is compiled, the IDE replaces any instance of the defined
word with the number that follows it. For example, when the IDE sees LED in
the line at ❷, it replaces it with the number 12.

We're basically using the #define command to label the digital pins for
the LED and button in the sketch. Also notice that we do not use a semicolon after a #define value. It's a good idea to label pin numbers and other
fixed values (such as a time delay) in this way, because if the value is used
repeatedly in the sketch, then you won't have to edit the same item more
than once. In this example, LED is used three times in the sketch; to edit this
value, we'd simply have to change its definition once in its #define statement.

Reading Digital Input Pins

To read the status of a button, we first define a digital I/O pin as an input
in void setup() using the following:

```
    pinMode(BUTTON, INPUT); // input for button
```

Next, to discover whether the button is connecting a voltage through to
the digital input (that is, it's being pressed), we use digitalRead(*pin*), where
pin is the digital pin number to read. The function returns either HIGH (voltage is close to 5 V at the pin) or LOW (voltage is close to 0 V at the pin).

Making Decisions with if

Using if, we can make decisions in our sketch and tell the Arduino to run different code, depending on the decision. For example, in the sketch for Project 4, we used Listing 4-2:

```
// Listing 4-2
if (digitalRead(BUTTON) == HIGH)
{
    digitalWrite(LED, HIGH);    // turn on the LED
    delay(500);                 // wait for 0.5 seconds
    digitalWrite(LED, LOW);     // turn off the LED
}
```

Listing 4-2: A simple if-then example

The first line in the code begins with if tests for a condition. If the condition is true (that is, voltage is HIGH), then it means that the button is pressed and the code that follows inside the curly brackets will run.

To determine whether the button is pressed (digitalRead(BUTTON) is set to HIGH), we use a *comparison operator,* a double equal sign (==). If we were to replace == with != (not equal to) in the sketch, then the LED would turn off when the button is pressed instead. Try it and see.

NOTE   *A common mistake is to use a single equal sign (=), which means "make equal to," in a test statement instead of a double equal sign (==), which says "test if it is equal to." You may not get an error message, but your if statement may not work!*

Making More Decisions with if-then-else

You can add another action to an if statement by using else. For example, if we rewrite Listing 4-1 by adding else as shown in Listing 4-3, then the LED will turn on *if* the button is pressed, or *else* it will be off. Using else forces the Arduino to run another section of code if the test in the if statement is not true.

```
// Listing 4-3
#define LED    12
#define BUTTON 7

void setup()
{
  pinMode(LED, OUTPUT);    // output for the LED
  pinMode(BUTTON, INPUT); // input for the button
}

void loop()
{
```

```
if ( digitalRead(BUTTON) == HIGH )
{
  digitalWrite(LED, HIGH);
}
else
{
  digitalWrite(LED, LOW);
}
}
```

Listing 4-3: Adding else

Boolean Variables

Sometimes you need to record whether something is in either of only two states, such as on or off, or hot or cold. A *Boolean variable* is the legendary computer "bit" whose value can be only a zero (0, false) or one (1, true). This is where the Boolean variable is useful: It can only be true or false. Like any other variable, we need to declare it in order to use it:

```
boolean raining = true; // create the variable "raining" and first make it true
```

Within the sketch, you can change the state of a Boolean with a simple reassignment, such as this:

```
raining = false;
```

It's simple to use Boolean variables to make decisions using an if test structure. Because Boolean comparisons can either be true or false, they work well with the comparison operators != and ==. Here's an example:

```
if ( raining == true )
{
    if ( summer != true )
    {
        // it is raining and not summer
    }
}
```

Comparison Operators

We can use various operators to make decisions about two or more Boolean variables or other states. These include the operators *not* (!), *and* (&&), and *or* (||).

The not Operator

The *not* operator is denoted by the use of an exclamation mark (!). This operator is used as an abbreviation for checking whether something is *not true*. Here's an example:

```
if ( !raining  )
{
    // it is not raining (raining == false)
}
```

The and Operator

The logical *and* operator is denoted by &&. Using *and* helps reduce the number of separate if tests. Here's an example:

```
if (( raining == true ) && ( !summer ))
{
    // it is raining and not summer (raining == true and summer == false)
}
```

The or Operator

The logical *or* operator is denoted by ||. Using *or* is very simple; here's an example:

```
if (( raining == true ) || ( summer == true ))
{
    // it is either raining or summer
}
```

Making Two or More Comparisons

You can also use two or more comparisons in the same if. Here's an example:

```
if ( snow == true && rain == true && !hot )
{
    // it is snowing and raining and not hot
}
```

And you can use parentheses to set the orders of operation. In the next example, the comparison in the parentheses is checked first, given a true or false state, and then compared with the rest in the if-then statement.

```
if (( snow == true || rain == true ) && hot == false))
{
    // it is either snowing or raining and not hot
}
```

Lastly, just like the examples of the not (!) operator before a value, simple tests of true or false can be performed without requiring == true or == false in each test. The following code works out the same as in the preceding example:

```
if (( snow || rain ) && !hot )
{
    // it is either snowing or raining and not hot
    // ( snow is true OR rain is true ) AND it is not hot
}
```

As you can see, it's possible to have the Arduino make a multitude of decisions using Boolean variables and comparison operators. Once you move on to more complex projects, this will become very useful.

Project #5: Controlling Traffic

Now let's put your newfound knowledge to use by solving a hypothetical problem. As the town planner for a rural shire, you have a problem with a single-lane bridge that crosses the river. Every week, one or two accidents occur at night, when tired drivers rush across the bridge without first stopping to see if the road is clear. You have suggested that traffic lights be installed, but the mayor wants to see them demonstrated before signing off on the purchase. You could rent temporary lights, but they're expensive. Instead, you've decided to build a model of the bridge with working traffic lights using LEDs and an Arduino.

The Goal

Our goal is to install three-color traffic lights at each end of the single-lane bridge. The lights allow traffic to flow only in one direction at a time. When sensors located at either end of the bridge detect a car waiting at a red light, the lights will change and allow the traffic to flow in the opposite direction.

The Algorithm

We'll use two buttons to simulate the vehicle sensors at each end of the bridge. Each set of lights will have red, yellow, and green LEDs. Initially, the system will allow traffic to flow from west to east, so the west-facing lights will be set to green and the east-facing lights will be set to red.

When a vehicle approaches the bridge (modeled by pressing the button) and the light is red, the system will turn the light on the opposite side from green to yellow to red, and then wait a set period of time to allow any vehicles already on the bridge to finish crossing. Next, the yellow light on the waiting vehicle's side will blink as a "get ready" notice for the driver, and finally the light will change to green. The light will remain green until a vehicle approaches the other side, at which point the process repeats.

The Hardware

Here's what you'll need to create this project:

- Two red LEDs (LED1 and LED2)
- Two yellow LEDs (LED3 and LED4)
- Two green LEDs (LED5 and LED6)
- Six 560 Ω resistors (R1 to R6)
- Two 10 kΩ resistor (R7 and R8)
- Two 100 nF capacitors (C1 and C2)
- Two push buttons (S1 and S2)
- One medium-sized breadboard
- One Arduino and USB cable
- Various connecting wires

The Schematic

Because we're controlling only six LEDs and receiving input from two buttons, the design will not be too difficult. Figure 4-26 shows the schematic for our project.

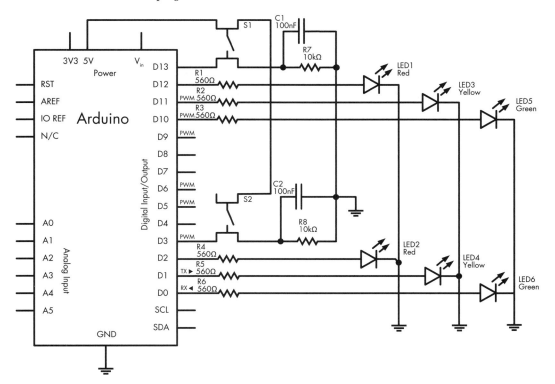

Figure 4-26: Schematic for Project 5

This circuit is basically a more elaborate version of the button and LED circuit in Project 4, with resistors, more LEDs, and another button.

Be sure that the LEDs are inserted in the correct direction: the resistors connect to LED anodes, and the LED cathodes connect to the Arduino GND pin, as shown in Figure 4-27.

Figure 4-27: Completed circuit

The Sketch

And now for the sketch. Can you see how it matches our algorithm?

```
// Project 5 - Controlling Traffic

// define the pins that the buttons and lights are connected to:
❶ #define westButton 3
#define eastButton 13
#define westRed     2
#define westYellow  1
#define westGreen   0
#define eastRed     12
#define eastYellow  11
#define eastGreen   10

#define yellowBlinkTime 500 // 0.5 seconds for yellow light blink
```

```
❷ boolean trafficWest = true;    // west = true, east = false
❸ int     flowTime    = 10000;   // amount of time to let traffic flow
❹ int     changeDelay = 2000;    // amount of time between color changes

void setup()
{
  // setup digital I/O pins
  pinMode(westButton, INPUT);
  pinMode(eastButton, INPUT);
  pinMode(westRed,    OUTPUT);
  pinMode(westYellow, OUTPUT);
  pinMode(westGreen,  OUTPUT);
  pinMode(eastRed,    OUTPUT);
  pinMode(eastYellow, OUTPUT);
  pinMode(eastGreen,  OUTPUT);

  // set initial state for lights - west side is green first
  digitalWrite(westRed,    LOW);
  digitalWrite(westYellow, LOW);
  digitalWrite(westGreen,  HIGH);
  digitalWrite(eastRed,    HIGH);
  digitalWrite(eastYellow, LOW);
  digitalWrite(eastGreen,  LOW);
}

void loop()
{
  if ( digitalRead(westButton) == HIGH ) // request west>east traffic flow
  {
    if ( trafficWest != true )
// only continue if traffic flowing in the opposite (east) direction
    {
      trafficWest = true; // change traffic flow flag to west>east
      delay(flowTime);    // give time for traffic to flow
      digitalWrite(eastGreen, LOW); // change east-facing lights from green
                                    // to yellow to red
      digitalWrite(eastYellow, HIGH);
      delay(changeDelay);
      digitalWrite(eastYellow, LOW);
      digitalWrite(eastRed, HIGH);
      delay(changeDelay);
      for ( int a = 0; a < 5; a++ ) // blink yellow light
      {
        digitalWrite(westYellow, LOW);
        delay(yellowBlinkTime);
        digitalWrite(westYellow, HIGH);
        delay(yellowBlinkTime);
      }
      digitalWrite(westYellow, LOW);
      digitalWrite(westRed, LOW); // change west-facing lights from red to green
      digitalWrite(westGreen, HIGH);
    }
  }
```

```
    if ( digitalRead(eastButton) == HIGH ) // request east>west traffic flow
    {
      if ( trafficWest == true )
// only continue if traffic flow is in the opposite (west) direction
      {
        trafficWest = false; // change traffic flow flag to east>west
        delay(flowTime);     // give time for traffic to flow
        digitalWrite(westGreen, LOW);
// change west lights from green to yellow to red
        digitalWrite(westYellow, HIGH);
        delay(changeDelay);
        digitalWrite(westYellow, LOW);
        digitalWrite(westRed, HIGH);
        delay(changeDelay);
        for ( int a = 0 ; a < 5 ; a++ ) // blink yellow light
        {
          digitalWrite(eastYellow, LOW);
          delay(yellowBlinkTime);
          digitalWrite(eastYellow, HIGH);
          delay(yellowBlinkTime);
        }
        digitalWrite(eastYellow, LOW);
        digitalWrite(eastRed, LOW); // change east-facing lights from red to green
        digitalWrite(eastGreen, HIGH);
      }
    }
}
```

Our sketch starts by using #define at ❶ to associate digital pin numbers with labels for all the LEDs used, as well as the two buttons. We have red, yellow, and green LEDs and a button each for the west and east sides of the bridge. The Boolean variable trafficWest at ❷ is used to keep track of which way the traffic is flowing—true is west to east, and false is east to west.

NOTE   *Notice that trafficWest is a single Boolean variable with the traffic direction set as either* true *or* false. *Having a single variable like this instead of two (one for east and one for west) ensures that both directions cannot accidentally be true at the same time, which helps avoid a crash!*

The integer variable flowTime at ❸ is the minimum period of time that vehicles have to cross the bridge. When a vehicle pulls up at a red light, the system delays this period to give the opposing traffic time to cross the bridge. The integer variable changeDelay at ❹ is the period of time between the traffic lights switching from green to yellow to red.

Before the sketch enters the void loop() section, it is set for traffic to flow from west to east in void setup().

Running the Sketch

Once it's running, the sketch does nothing until one of the buttons is pressed. When the east button is pressed, the line

```
if ( trafficWest == true )
```

ensures that the lights change only if the traffic is heading in the opposite direction. The rest of the code section is composed of a simple sequence of waiting and then of turning on and off various LEDs to simulate the traffic-light operation.

Analog vs. Digital Signals

In this section, you'll learn the difference between digital and analog signals, and you'll learn how to measure analog signals with the analog input pins.

Until now, our sketches have been using digital electrical signals, with just two discrete levels. Specifically, we used `digitalWrite(pin, HIGH)` and `digitalWrite(pin, LOW)` to blink an LED and `digitalRead()` to measure whether a digital pin had a voltage applied to it (`HIGH`) or not (`LOW`). Figure 4-28 is a visual representation of a digital signal that alternates between high and low.

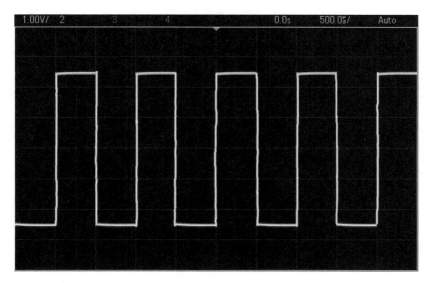

Figure 4-28: A digital signal, with HIGHs appearing as horizontal lines at the top, and LOWs appearing at the bottom

Unlike digital signals, analog signals can vary with an indefinite number of steps between high and low. For example, Figure 4-29 shows an analog signal of a sine wave. Notice in the figure that as time progresses, the voltage moves fluidly between high and low levels.

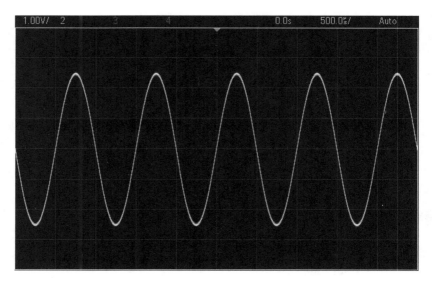

Figure 4-29: An analog signal of a sine wave

With our Arduino, high is closer to 5 V and low is closer to 0 V, or GND. We can measure the voltage values of an analog signal with our Arduino using the six analog inputs shown in Figure 4-30. These analog inputs can safely measure voltages from 0 (GND) to no more than 5 V.

If you use the function analogRead(), then the Arduino will return a number between 0 and 1,023 in proportion to the voltage applied to the analog pin. For example, you might use analogRead() to store the value of analog pin zero in the integer variable a:

Figure 4-30: Analog inputs on the Arduino Uno

```
a = analogRead(0); // read analog input pin 0 (A0)
                   // returns 0 to 1023 which is usually 0.000 to 4.995
volts
```

Project #6: Creating a Single-Cell Battery Tester

Although the popularity and use of cell batteries has declined, most people still have a few devices around the house that use AA, AAA, C, or D cell batteries, such as remote controls, clocks, or children's toys. These batteries carry much less than 5 V, so we can measure a cell's voltage with our Arduino to determine the state of the cell. In this project we'll create a battery tester.

The Goal

Single-cell batteries such as AAs usually begin at about 1.6 V when new and then decrease with use. We will measure the voltage and express the battery condition visually with LEDs. We'll use the reading from `analogRead()` and then convert the reading to volts. The maximum voltage that can be read is 5 V, so we divide 5 by 1,024 (the number of possible values), which equals 0.0048. Therefore, if `analogRead()` returns 512, then we multiply that reading by 0.0048, which equals 2.4576 V.

The Algorithm

Here's the algorithm for our battery tester operation:

1. Read from analog pin zero.
2. Multiply the reading by 0.0048 to create a voltage value.
3. If the voltage is greater than or equal to 1.6 V, then briefly turn on a green LED.
4. If the voltage is greater than 1.4 V *and* less than 1.6 V, then briefly turn on a yellow LED.
5. If the voltage is less than 1.4 V, then briefly turn on a red LED.
6. Repeat indefinitely.

The Hardware

Here's what you'll need to create this project:

*   Three 560 Ω resistors (R1 to R3)
*   One green LED (LED1)
*   One yellow LED (LED2)
*   One red LED (LED3)
*   One breadboard
*   Various connecting wires
*   One Arduino and USB cable

The Schematic

The schematic for the single-cell battery tester circuit is shown in Figure 4-31. On the left side, notice the two terminals, labeled + and −. Connect the *matching* sides of the single-cell battery to be tested at those points. Positive should connect to positive, and negative should connect to negative.

Under no circumstances should you measure anything larger than 5 V, nor should you connect positive to negative, or vice versa. Doing these things will damage your Arduino board.

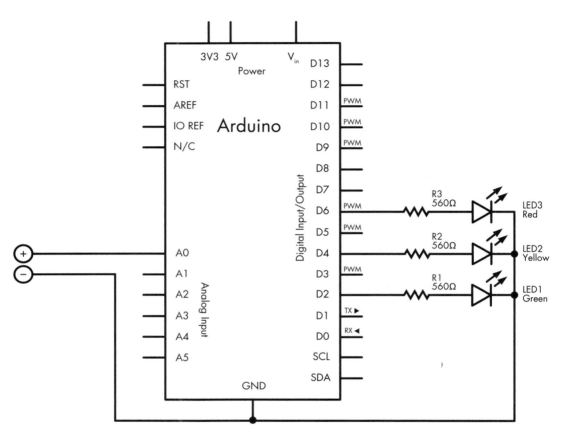

Figure 4-31: Schematic for Project 6

The Sketch

Now for the sketch:

```
// Project 6 - Creating a Single-Cell Battery Tester
#define newLED 2  // green LED  'new'
#define okLED  4  // yellow LED 'ok'
#define oldLED 6  // red LED    'old'

int analogValue = 0;
float voltage = 0;
int ledDelay = 2000;
```

❶

```
void setup()
{
  pinMode(newLED, OUTPUT);
  pinMode(okLED, OUTPUT);
  pinMode(oldLED, OUTPUT);
}

void loop()
{
❷  analogValue = analogRead(0);
❸  voltage = 0.0048*analogValue;
❹  if ( voltage >= 1.6 )
  {
    digitalWrite(newLED, HIGH);
    delay(ledDelay);
    digitalWrite(newLED, LOW);
  }
❺  else if ( voltage < 1.6 && voltage > 1.4 )
  {
    digitalWrite(okLED, HIGH);
    delay(ledDelay);
    digitalWrite(okLED, LOW);
  }
❻  else if ( voltage <= 1.4 )
  {
    digitalWrite(oldLED, HIGH);
    delay(ledDelay);
    digitalWrite(oldLED, LOW);
  }
}
```

In the sketch for Project 6, the Arduino takes the value measured by analog pin 0 at ❷ and converts this to a voltage at ❸. You'll learn about a new type of variable, float at ❶, in the next section. You'll also see some familiar code, such as the if-else functions, and some new topics, such as doing arithmetic and using comparison operators to compare numbers, which are all discussed in the sections that follow.

Doing Arithmetic with an Arduino

Like a pocket calculator, the Arduino can perform calculations for us, such as multiplication, division, addition, and subtraction. Here are some examples:

```
a = 100;
b = a + 20;
c = b - 200;
d = c + 80; // d will equal 0
```

Float Variables

When you need to deal with numbers with a decimal point, you can use the variable type float. The values that can be stored in a float fall between 3.4028235 × 1038 and −3.4028235 × 1038, and are generally limited to six or seven decimal places of precision. And you can mix integers and float numbers in your calculations. For example, you could add the float number f to the integer a then store it as the float variable g:

```
int a = 100;
float f;
float g;

    f = a / 3; // f = 33.333333
    g = a + f; // g = 133.333333
```

Comparison Operators for Calculations

We used comparison operators such as == and != with if statements and digital input signals in Project 5. In addition to these operators, we can also use the following to compare numbers or numerical variables:

- <   less than
- >   greater than
- <= less than or equal to
- >= greater than or equal to

We've used these comparison operators to compare numbers in lines ❹, ❺, and ❻ in the sketch for Project 6 described earlier.

Improving Analog Measurement Precision with a Reference Voltage

As demonstrated in Project 6, the analogRead() function returns a value proportional to a voltage between 0 and 5 V. The upper value (5 V) is the *reference voltage*, the maximum voltage that the Arduino analog inputs will accept and return the highest value for (1,023).

To increase precision while reading even lower voltages, we can use a lower reference voltage. For example, when the reference voltage is 5 V, analogRead() represents this with a value from 0 to 1,023. However, if we need to measure only a voltage with a maximum of (for example) 2 V, then we can alter the Arduino output to represent 2 V using the 0–1,023 value range to allow for more precise measurement. You can do this with either an external or internal reference voltage, as discussed next.

Using an External Reference Voltage

The first method of using a reference voltage is with the *AREF* (*a*nalog *ref*erence) pin, as shown in Figure 4-32.

Figure 4-32: The Arduino Uno AREF pin

We can introduce a new reference voltage by connecting the voltage into the AREF pin and the matching GND to the Arduino's GND. Note that this can lower the reference voltage but will not raise it, because the reference voltage connected to an Arduino Uno must not exceed 5 V. A simple way to set a lower reference voltage is by creating a *voltage divider* with two resistors, as shown in Figure 4-33.

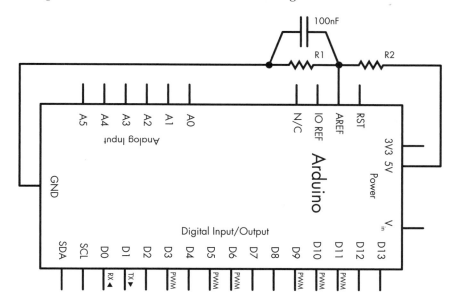

Figure 4-33: Voltage divider circuit

The values of *R1* and *R2* will determine the reference voltage according to the following formula:

$$V_{out} = V_{in} \left(\frac{R2}{R1+R2} \right)$$

*V*_{*out*} is the reference voltage, and *V*_{*in*} is the input voltage—in this case 5 V. *R1* and *R2* are the resistor values in ohms.

The simplest way to divide the voltage is to split V_{in} in half by setting *R1* and *R2* to the same value—for example, 10 kΩ each. When you're doing this, it's best to use the lowest-tolerance resistors you can find, such as 1 percent; confirm their true resistance values with a multimeter, and use those confirmed values in the calculation. Furthermore, it's also a very good idea to place a 100 nF capacitor between AREF and GND to avoid a noisy AREF and prevent unstable analog readings.

When using an external reference voltage, insert the following line in the void setup() section of your sketch:

```
analogReference(EXTERNAL); // select AREF pin for reference voltage
```

Using the Internal Reference Voltage

The Arduino Uno also has an internal 1.1 V reference voltage. If this meets your needs, no hardware changes are required. Just add this line to void setup():

```
analogReference(INTERNAL); // select internal 1.1 V reference voltage
```

The Variable Resistor

Variable resistors, also known as *potentiometers*, can generally be adjusted from 0 Ω up to their rated value. Their schematic symbol is shown in Figure 4-34.

Variable resistors have three pin connections: one in the center pin and one on each side. As the shaft of a variable resistor turns, it increases the resistance between one side and the center and decreases the resistance between the center and the opposite side.

Variable resistors are available as *linear* and *logarithmic*. The resistance of linear models changes at a constant rate when turning, while the resistance of logarithmic models changes slowly at first and then increases rapidly. Logarithmic potentiometers are used more often in audio amplifier circuits, because they model the human hearing response. Most Arduino projects use linear variable resistors such as the one shown in Figure 4-35.

Figure 4-34: Variable resistor (potentiometer) symbol

Figure 4-35: A typical linear variable resistor

You can also get miniature versions of variable resistors, known as *trimpots* or *trimmers* (see Figure 4-36). Because of their size, trimpots are more useful for making adjustments in circuits, but they're also very useful for breadboard work because they can be slotted in.

Figure 4-36: Various trimpots

NOTE *When shopping for trimpots, take note of the type. Often you will want one that is easy to adjust with a screwdriver that you have on hand, and the enclosed types, as pictured in Figure 4-36, last longer than the cheaper, open contact types.*

Piezoelectric Buzzers

A *piezoelectric element* (*piezo* for short), or buzzer, is a small, round device that can be used to generate loud and annoying noises that are perfect for alarms—or for having fun. Figure 4-37 shows a common example, the TDK PS1240, next to an American quarter, to give you an idea of its size.

Figure 4-37: TDK PS1240 Piezo

Piezos contain a very thin plate inside the housing that moves when an electrical current is applied. When a pulsed current is applied (such as on . . . off . . . on . . . off), the plate vibrates and generates sound waves.

It's simple to use piezos with Arduino because they can be turned on and off just like an LED. The piezo elements are not polarized and can be connected in either direction.

Piezo Schematic

The schematic symbol for the piezo looks like a loudspeaker (Figure 4-38), which makes it easy to recognize.

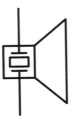

Figure 4-38: Piezo schematic

NOTE   *When shopping for a piezo for this project, be sure to get the* piezo element only *type; some buzzer types look like Figure 4-38 but include a tone-generating circuit built into the case. We don't want those because we're going to drive our tone directly from the Arduino.*

Project #7: Trying Out a Piezo Buzzer

If you have a piezo handy and want to try it out, upload the following demonstration sketch to your Arduino:

```
// Project 7 - Trying Out a Piezo Buzzer
#define PIEZO 3  // pin 3 is capable of PWM output to drive tones
int del = 500;
void setup()
{
  pinMode(PIEZO, OUTPUT);
}

void loop()
{
  analogWrite(PIEZO, 128);  // 50 percent duty cycle tone to the piezo
  delay(del);
  digitalWrite(PIEZO, LOW); // turn the piezo off
  delay(del);
}
```

❶

This sketch uses pulse-width modulation on digital pin three. If you change the duty cycle in the analogWrite() function (currently it's 128, which is 50 percent on) at ❶, then you can alter the volume of the buzzer.

To increase the volume of your piezo, increase the voltage applied to it. The voltage is currently limited to 5 V, but the buzzer would be much louder at 9 or 12 V. Because higher voltages can't be sourced from the Arduino, you would need to use an external power source for the buzzer, such as a 9 V battery, and then switch the power into the buzzer using a transistor as an electronic switch. You can use the same sketch with the schematic shown in Figure 4-39.

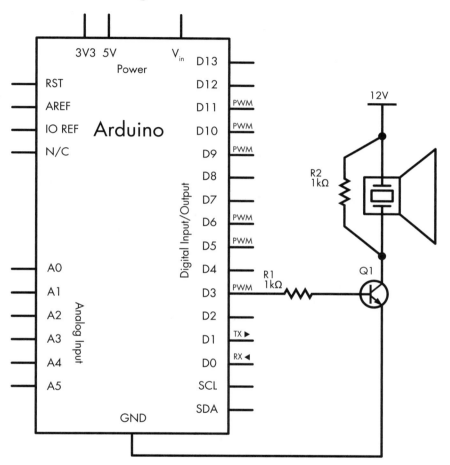

Figure 4-39: Schematic for Project 7

The part of the schematic labeled 12 V will be the positive side of the higher-power supply, whose negative side will connect to the Arduino GND pin.

Temperature can be represented by an analog signal. We can measure temperature using the TMP36 voltage output temperature sensor made by Analog Devices (*http://www.analog.com/tmp36/*), shown in Figure 4-40.

Notice that the TMP36 looks just like the BC548 transistor we worked with in the relay control circuit in Chapter 3. The TMP36 outputs a voltage that is proportional to the temperature, so you can determine the current temperature using a simple conversion. For example, at 25 degrees Celsius, the output voltage is 750 mV, and each change in temperature of 1 degree results in a change of 10 mV. The TMP36 can measure temperatures between −40 and 125 degrees Celsius.

Figure 4-40: TMP36 temperature sensor

The function analogRead() will return a value between 0 and 1,023, which corresponds to a voltage between 0 and just under 5,000 mV (5 V). If we multiply the output of analogRead() by (5,000/1,024), then we will get the actual voltage returned by the sensor. Next, we subtract 500 (an offset used by the TMP36 to allow for temperatures below zero) and then divide by 10, which leaves us with the temperature in degrees Celsius. If you work in Fahrenheit, then multiply the Celsius value by 1.8 and add 32 to the result.

The Goal

In this project, we'll use the TMP36 to create a quick-read thermometer. When the temperature falls below 20 degrees Celsius, a blue LED turns on. When the temperature is between 20 and 26 degrees, a green LED turns on, and when the temperature is above 26 degrees, a red LED turns on.

The Hardware

Here's what you'll need to create this project:

- Three 560 Ω resistors (R1 to R3)
- One red LED (LED1)
- One green LED (LED2)
- One blue LED (LED3)
- One TMP36 temperature sensor
- One breadboard
- Various connecting wires
- Arduino and USB cable

The Schematic

The circuit is simple. When you're looking at the labeled side of the TMP36, the pin on the left connects to the 5 V input, the center pin is the voltage output, and the pin on the right connects to GND as shown in Figure 4-41.

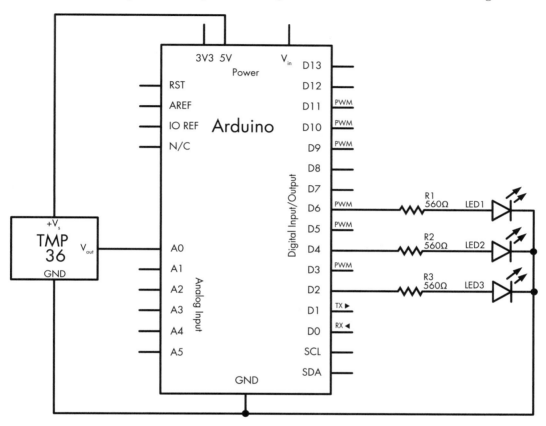

Figure 4-41: Schematic for Project 8

The Sketch

And now for the sketch:

```
// Project 8 - Creating a Quick-Read Thermometer

// define the pins that the LEDs are connected to:
#define HOT     6
#define NORMAL  4
#define COLD    2

float voltage = 0;
float celsius = 0;
```

```
float hotTemp  = 26;
float coldTemp = 20;
float sensor = 0;

void setup()
{
  pinMode(HOT, OUTPUT);
  pinMode(NORMAL, OUTPUT);
  pinMode(COLD, OUTPUT);
}

void loop()
{
    // read the temperature sensor and convert the result to degrees Celsius
❶  sensor = analogRead(0);
    voltage = (sensor*5000)/1024; // convert raw sensor value to millivolts
    voltage = voltage-500;        // remove voltage offset
    celsius = voltage/10;         // convert millivolts to Celsius

    // act on temperature range
❷  if ( celsius < coldTemp )
    {
      digitalWrite(COLD, HIGH);
      delay(1000);
      digitalWrite(COLD, LOW);
    }
❸  else if ( celsius > coldTemp && celsius <= hotTemp )
    {
      digitalWrite(NORMAL, HIGH);
      delay(1000);
      digitalWrite(NORMAL, LOW);
    }
    else
    {
      // celsius is > hotTemp
      digitalWrite(HOT, HIGH);
      delay(1000);
      digitalWrite(HOT, LOW);
    }
}
```

The sketch first reads the voltage from the TMP36 and converts it to temperature in degrees Celsius at ❶. Next, using the if-else functions at ❷ and ❸, the code compares the current temperature against the values for hot and cold and turns on the appropriate LED. The delay(1000) statements are used to prevent the lights from flashing on and off too quickly if the temperature fluctuates rapidly between two ranges.

Hacking the Sketch

Although this sketch was rather simple, you could use it as the basis for taking other sorts of readings. You might add a PowerSwitch Tail, for example, as shown in Figure 4-42.

Figure 4-42: A PowerSwitch Tail that switches up to 120 V AC

With a PowerSwitch Tail, you can safely control an appliance that runs from the wall socket, such as a heater, lamp, or another device with a digital output from your Arduino. (For more information, visit *http://www.adafruit.com/products/268/*.) For example, you could use a PowerSwitch Tail to build a temperature-controlled heater or fan, control a garage light so it runs for a time and then switches off, or remotely control outdoor Christmas lights.

Looking Ahead

And Chapter 4 comes to a close. You now have a lot more tools to work with, including digital inputs and outputs, new types of variables, and various mathematical functions. In the next chapter, you will have a lot more fun with LEDs, learn to create your own functions, build a computer game and electronic dice, and much more.

5

WORKING WITH FUNCTIONS

In this chapter you will

- Create your own functions
- Learn to make decisions with `while` and `do-while`
- Send and receive data between your Arduino and the Serial Monitor window
- Learn about `long` variables

You'll learn new methods to make your Arduino sketches easier to read and simpler to design by creating your own functions. You can also create modular, reusable code that will save you time again and again. We'll introduce a way to make decisions that control blocks of code, and you'll learn about a type of integer variable called the `long`. Then you will use your own functions to create a new type of thermometer.

A *function* consists of a set of instructions that we can use anywhere in our sketches. Although many functions are available in the Arduino language, sometimes you won't find one to suit your specific needs—or you may need to run part of a sketch repeatedly to make it work, which is a waste of memory. In both of these situations, you might wish you had a better function to do what you need to do. The good news is that there is such a function—the one you create yourself.

Project #9: Creating a Function to Repeat an Action

You can write simple functions to repeat actions on demand. For example, the following function will turn the built-in LED on (at ❶ and ❸) and off (at ❷ and ❹) twice.

```
void blinkLED()
{
❶      digitalWrite(13, HIGH);
        delay(1000);
❷      digitalWrite(13, LOW);
        delay(1000);
❸      digitalWrite(13, HIGH);
        delay(1000);
❹      digitalWrite(13, LOW);
        delay(1000);
}
```

Here is the function being used within a complete sketch, which you can upload to the Arduino:

```
// Project 9 - Creating a Function to Repeat an Action

#define LED 13
#define del 200

void setup()
{
  pinMode(LED, OUTPUT);
}

void blinkLED()
{
  digitalWrite(LED, HIGH);
  delay(del);
  digitalWrite(LED, LOW);
  delay(del);
  digitalWrite(LED, HIGH);
  delay(del);
  digitalWrite(LED, LOW);
  delay(del);
}

void loop()
{
❶  blinkLED();
  delay(1000);
}
```

When the blinkLED() function is called in void loop() at ❶, the Arduino will run the commands within the void blinkLED() section. In other words, you have created your own function and used it when necessary.

Project #10: Creating a Function to Set the Number of Blinks

The function we just created is pretty limited. What if we want to set the number of blinks and the delay? No problem; we can create a function that lets us change values, like this:

```
void blinkLED(int cycles, int del)
{
  for ( int z = 0 ; z < cycles ; z++ )
  {
    digitalWrite(LED, HIGH);
    delay(del);
    digitalWrite(LED, LOW);
    delay(del);
  }
}
```

Our new `void blinkLED()` function accepts two integer values: cycles (the number of times we want to blink the LED) and del (the delay time between turning the LED on and off). So if we wanted to blink the LED 12 times, with a 100-millisecond delay, then we would use `blinkLED(12, 100)`. Enter the following sketch into the IDE to experiment with this function:

```
// Project 10 - Creating a Function to Set the Number of Blinks

#define LED 13

void setup()
{
  pinMode(LED, OUTPUT);
}

void blinkLED(int cycles, int del)
{
  for ( int z = 0 ; z < cycles ; z++ )
  {
    digitalWrite(LED, HIGH);
    delay(del);
    digitalWrite(LED, LOW);
    delay(del);
  }
}

void loop()
{
❶  blinkLED(12, 100);
  delay(1000);
}
```

You can see at ❶ that the values of 12 and 100 (for the number of blinks and the delay, respectively) are passed into our custom function blinkLED(), where cycles will have a value of 12 and del will have a value of 100. Therefore, the LED will blink 12 times with a delay of 100 milliseconds between blinks.

Creating a Function to Return a Value

In addition to creating functions that accept values entered as parameters (as void blinkLED() did in Project 10), you can also create functions that return a value, in the same way that analogRead() returns a value between 0 and 1,023 when measuring an analog input, as demonstrated in Project 8. The void that appears at the start of functions up to this point means that the function doesn't return anything—that is, the function's return value is void. Let's create some useful functions that return actual values.

Consider this function that converts degrees Celsius to Fahrenheit:

```
float convertTemp(float celsius)
{
  float fahrenheit = 0;
  fahrenheit = (1.8 * celsius) + 32;
  return fahrenheit;
}
```

In the first line, we define the function name (convertTemp), its return variable type (float), and any variables that we might want to pass into the function (float celsius). To use this function, we send it an existing variable. For example, if we wanted to convert 40 degrees Celsius to Fahrenheit and store the result in a float variable called tempf, then we would call convertTemp like so:

```
tempf = convertTemp(40);
```

This would place 40 into the convertTemp variable celsius and use it in the calculation fahrenheit = (1.8 * celsius) + 32 in the convertTemp function. The result is then returned into the variable tempf with the convertTemp line return fahrenheit.

Project #11: Creating a Quick-Read Thermometer That Blinks the Temperature

Now that you know how to create custom functions, we'll make a quick-read thermometer using the TMP36 temperature sensor from Chapter 4 and the Arduino's built-in LED. If the temperature is below 20 degrees Celsius, the LED will blink twice and then pause; if the temperature falls between 20 and 26 degrees, the LED will blink four times and then pause; and if the temperature is above 26 degrees, the LED will blink six times.

We'll make our sketch more modular by breaking it up into distinct functions that will make the sketch easier to follow, and the functions will be reusable. Our thermometer will perform two main tasks: measure and categorize the temperature, and blink the LED a certain number of times (determined by the temperature).

The Hardware

The required hardware is minimal:

- One TMP36 temperature sensor
- One breadboard
- Various connecting wires
- Arduino and USB cable

The Schematic

The circuit is very simple, as shown in Figure 5-1.

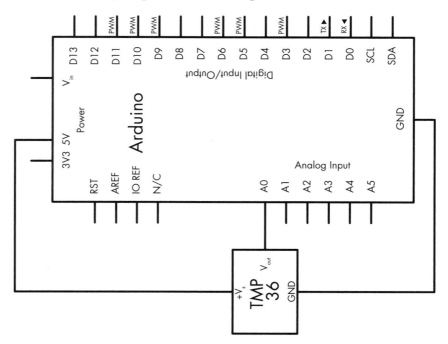

Figure 5-1: Schematic for Project 11

The Sketch

We'll need to create two functions for the sketch. The first one will read the value from the TMP36, convert it to Celsius, and then return a value of *2*, *4*, or *6*, corresponding to the number of times the LED should blink. We'll alter the sketch from Project 8 for this purpose.

For our second function, we'll use blinkLed() from Project 9. Our void loop will call the functions in order and then pause for 2 seconds before restarting.

NOTE   *Remember to save your modified project sketches with new filenames so that you don't accidentally delete your existing work!*

For the sketch, enter this code into the IDE:

```
// Project 11 - Creating a Quick-Read Thermometer That Blinks the Temperature

#define LED 13

int blinks = 0;

void setup()
{
  pinMode(LED, OUTPUT);
}

int checkTemp()
{
  float voltage  = 0;
  float celsius  = 0;
  float hotTemp  = 26;
  float coldTemp = 20;
  float sensor   = 0;
  int result;
  // read the temperature sensor and convert the result to degrees Celsius

  sensor = analogRead(0);
  voltage = (sensor * 5000) / 1024; // convert raw sensor value to millivolts
  voltage = voltage - 500;          // remove voltage offset
  celsius = voltage / 10;           // convert millivolts to Celsius

  // act on temperature range
  if (celsius < coldTemp)
  {
    result = 2;
  }
```

```
  else if (celsius >= coldTemp && celsius <= hotTemp)
  {
    result = 4;
  }
  else
  {
    result = 6;    // (celsius > hotTemp)
  }
  return result;
}

void blinkLED(int cycles, int del)
{
  for ( int z = 0 ; z < cycles ; z++ )
  {
    digitalWrite(LED, HIGH);
    delay(del);
    digitalWrite(LED, LOW);
    delay(del);
  }
}

void loop()
{
  blinks = checkTemp();
  blinkLED(blinks, 500);
  delay(2000);
}
```

Because we use custom functions, all we have to do in void_loop() is call them and set the delay. The function checkTemp() returns a value to the integer variable blinks, and then blinkLED() will blink the LED blinks times with a delay of 500 milliseconds. The sketch then pauses for 2 seconds before repeating.

Upload the sketch and watch the LED to see this thermometer in action. (Be sure to keep this circuit assembled, since we'll use it in the following examples.)

Displaying Data from the Arduino in the Serial Monitor

So far, we have sent sketches to the Arduino and used the LEDs to show us output (such as temperature and traffic signals). Blinking LEDs make it easy to get feedback from the Arduino, but blinking lights can tell us only so much. In this section you'll learn how to use the Arduino's cable connection and the IDE's Serial Monitor window to display data from the Arduino and send data to the Arduino from the computer keyboard.

The Serial Monitor

To open the Serial Monitor, start the IDE and click the Serial Monitor icon button on the tool bar, shown in Figure 5-2. The Serial Monitor should open and look similar to Figure 5-3.

Figure 5-2: Serial Monitor icon button on the IDE tool bar

Figure 5-3: Serial Monitor

As you can see in Figure 5-3, the Serial Monitor displays an input field at the top, consisting of a single row and a Send button, and an output window below it, where data from the Arduino is displayed. When the Autoscroll box is checked, the most recent output is displayed, and once the screen is full, older data rolls off the screen as newer output is received. If you uncheck Autoscroll, you can manually examine the data using a vertical scroll bar.

Starting the Serial Monitor

Before we can use the Serial Monitor, we need to activate it by adding this function to our sketch in `void setup()`:

```
Serial.begin(9600);
```

The value 9600 is the speed at which the data will travel between the computer and the Arduino, also known as *baud*. This value must match the speed setting at the bottom right of the Serial Monitor, as shown in Figure 5-3.

Sending Text to the Serial Monitor

To send text to the Serial Monitor to be displayed in the output window, you can use `Serial.print`:

```
Serial.print("Arduino for Everyone!");
```

This sends the text between the quotation marks to the Serial Monitor's output window.

You can also use `Serial.println` to display text and then force any following text to start on the next line:

```
Serial.println("Arduino for Everyone!");
```

Displaying the Contents of Variables

You can also display the contents of variables on the Serial Monitor. For example, this would display the contents of the variable results:

```
Serial.println(results);
```

If the variable is a `float`, the display will default to two decimal places. You can specify the number of decimal places used as a number between 0 and 6 by entering a second parameter after the variable name. For example, to display the `float` variable results to four decimal places, you would enter the following:

```
Serial.print(results,4);
```

Project #12: Displaying the Temperature in the Serial Monitor

Using the hardware from Project 8, we'll display temperature data in Celsius and Fahrenheit in the Serial Monitor window. To do this, we'll create one function to determine the temperature values and another to display them in the Serial Monitor.

Enter this code into the IDE:

```
// Project 12 - Displaying the Temperature in the Serial Monitor

float celsius    = 0;
float fahrenheit = 0;

void setup()
{
  Serial.begin(9600);
}

❶ void findTemps()
  {
    float voltage = 0;
    float sensor  = 0;
```

```
    // read the temperature sensor and convert the result to degrees C and F
    sensor  = analogRead(0);
    voltage = (sensor * 5000) / 1024;  // convert the raw sensor value to millivolts
    voltage = voltage - 500;           // remove the voltage offset
    celsius = voltage / 10;            // convert millivolts to Celsius
    fahrenheit = (1.8 * celsius) + 32; // convert Celsius to Fahrenheit
  }

❷ void displayTemps()
  {
     Serial.print("Temperature is ");
     Serial.print(celsius, 2);
     Serial.print(" deg. C / ");
     Serial.print(fahrenheit, 2);
     Serial.println(" deg. F");
  // use .println here so the next reading starts on a new line
  }

  void loop()
  {
    findTemps();
    displayTemps();
    delay(1000);
  }
```

A lot is happening in this sketch, but we've created two functions, findTemps() at ❶ and displayTemps() at ❷, to simplify things. These functions are called in void loop(), which is quite simple. Thus you see one reason to create your own functions: to make your sketches easier to understand and the code more modular and possibly reusable.

After uploading the sketch, wait a few seconds, and then display the Serial Monitor. The temperature in your area should be displayed in a similar manner to that shown in Figure 5-4.

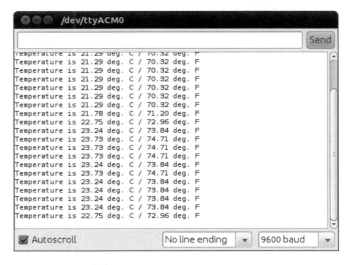

Figure 5-4: Result from Project 12

Debugging with the Serial Monitor

The Serial Monitor can be used to help *debug* (locate and fix errors) your sketch. For example, if you insert Serial.println(); statements in your sketch containing brief notes about the location in the sketch, then you can see when the Arduino passes each statement. For example, you might use the line

```
Serial.println("now in findTemps()");
```

inside the function findTemps() to let you know when the Arduino is running that particular function.

Making Decisions with while Statements

You can use while() statements in a sketch to repeat instructions, as long as (*while*) a given condition is true. The condition is always tested *before* the code in the while() statement is executed. For example, while (temperature > 30) will test to determine if the value of temperature is greater than 30. You can use any comparison operator within the parentheses to create the condition.

In the following listing, the Arduino will count up to 10 seconds and then continue with its program:

```
int a = 0; // an integer
while ( a < 10 )
{
    a = a + 1;
    delay(1000);
}
```

This sketch starts with the variable a set to 0. It then adds 1 to the value of a (which starts at 0), waits 1 second (delay(1000)), and then repeats the process until a has a value of 10 (while (a < 10)). Once a is equal to 10, the comparison in the while statement is false; therefore, the Arduino will continue on with the sketch after the while loop brackets.

do-while

In contrast to while, the do-while() structure places the test *after* the code within the do-while statement is executed. Here's an example:

```
int a = 0; // an integer
do
{
    delay(1000);
    a = a + 1;
} while ( a < 100 );
```

In this case, the code between the curly brackets will execute *before* the conditions of the test (while (a < 100)) have been checked. As a result, even if the conditions are not met, the loop will run once. You'll decide whether to use a while or a do-while function when designing your particular project.

Sending Data from the Serial Monitor to the Arduino

To send data from the Serial Monitor to the Arduino, we need the Arduino to listen to the *serial buffer*—the part of the Arduino that receives data from the outside world via the serial pins (digital 0 and 1) that are also connected to the USB circuit and cable to your computer. The serial buffer holds incoming data from the Serial Monitor's input window.

Project #13: Multiplying a Number by Two

To demonstrate the process of sending and receiving data via the Serial Monitor, let's dissect the following sketch. This sketch accepts a single digit from the user, multiplies it by 2, and then displays the result in the Serial Monitor's output window.

```
// Project 13 - Multiplying a Number by Two

int number;

void setup()
{
  Serial.begin(9600);
}
void loop()
{
  number = 0;      // zero the incoming number ready for a new read
  Serial.flush(); // clear any "junk" out of the serial buffer before waiting
❶ while (Serial.available() == 0)
  {
    // do nothing until something enters the serial buffer
  }
❷ while (Serial.available() > 0)
  {
    number = Serial.read() - '0';
// read the number in the serial buffer,
// remove the ASCII text offset for zero: '0'
  }
  // Show me the number!
  Serial.print("You entered: ");
  Serial.println(number);
  Serial.print(number);
  Serial.print(" multiplied by two is ");
  number = number * 2;
  Serial.println(number);
}
```

The `Serial.available()` test in the first `while` statement at ❶ returns 0 if nothing is entered yet into the Serial Monitor by the user. In other words, it tells the Arduino, "Do nothing until the user enters something." The next `while` statement at ❷ detects the number in the serial buffer and converts the text code that represents the data entered into an actual integer number. Afterward, the Arduino displays the number from the serial buffer and the multiplication results.

The `Serial.flush()` function at the start of the sketch clears the serial buffer just in case any unexpected data is in it, readying it to receive the next available data. Figure 5-5 shows the Serial Monitor window after the sketch has run.

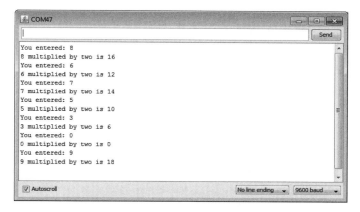

Figure 5-5: Sample input and output for Project 13

Although you can now enter numerical data into the Serial Monitor for the Arduino to process, using integer variables limits the range of numbers available. We can use `long` variables to increase this range, as discussed next.

long Variables

To use the Serial Monitor to accept numbers with more than one digit, we need to add some new code to our sketch, as you'll see shortly. When working with larger numbers, however, the `int` variable type can be limiting because it has a maximum value of 32,767. Fortunately, we can extend this limitation by using the `long` variable type. A `long` variable is a whole number between –2,147,483,648 and 2,147,483,647, a much larger range than that of an `int` variable (–32,768 to 32,767).

Project #14: Using long Variables

We'll use the Serial Monitor to accept `long` variables and numbers larger than one digit. This sketch accepts a number of many digits, multiplies that number by 2, and then returns the result to the Serial Monitor.

```
// Project 14 - Using long Variables

long number = 0;
long a = 0;

void setup()
{
  Serial.begin(9600);
}

void loop()
{
  number = 0;      // zero the incoming number ready for a new read
  Serial.flush(); // clear any "junk" out of the serial buffer before waiting
  while (Serial.available() == 0)
  {
    // do nothing until something comes into the serial buffer,
    // when something does come in, Serial.available will return how many
    // characters are waiting in the buffer to process
  }
  // one character of serial data is available, begin calculating
  while (Serial.available() > 0)
  {
    // move any previous digit to the next column on the left;
    // in other words, 1 becomes 10 while there is data in the buffer
    number = number * 10;
    // read the next number in the buffer and subtract the character 0
    // from it to convert it to the actual integer number
    a = Serial.read() - '0';
    // add this value a into the accumulating number
    number = number + a;
    // allow a short delay for more serial data to come into Serial.available
    delay(5);
  }
  Serial.print("You entered: ");
  Serial.println(number);
  Serial.print(number);
  Serial.print(" multiplied by two is ");
  number = number * 2;
  Serial.println(number);
}
```

In this example, two while loops allow the Arduino to accept multiple digits from the Serial Monitor. When the first digit is entered (the leftmost digit of the number entered), it is converted to a number and then added to the total variable number. If that's the only digit, the sketch moves on. If another digit is entered (for example, the *2* in *42*), then the total is multiplied by 10 to shift the first digit to the left, and then the new digit is added to the total. This cycle repeats until the rightmost digit has been added to the total. Don't forget to select **No line ending** in the Serial Monitor window.

Figure 5-6 shows the input and output of this sketch.

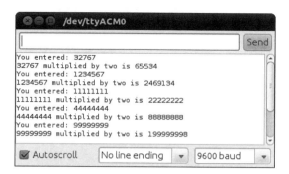

Figure 5-6: Sample input and output from Project 14

Looking Ahead

Although this chapter may have seemed a little dry, the ability to create your own functions is an important skill that will simplify your sketches and save time and effort. You will make good use of this knowledge in the next chapter.

6

NUMBERS, VARIABLES, AND ARITHMETIC

In this chapter you will

- Generate random numbers
- Create electronic dice
- Learn about binary numbers
- Use shift-register integrated circuits (ICs) to get more digital output pins
- Test your knowledge of binary numbers with a quiz
- Learn about arrays of variables
- Display numbers on seven-segment LED modules
- Learn how to use the modulo math function
- Create a digital thermometer
- Learn about bitwise arithmetic
- Create fixed and moving images on LED matrix displays

You will learn a wide variety of useful new functions that will create more project options, including random number generation, new kinds of math functions, and variable storage in ordered lists called *arrays*. Furthermore, you will learn how to use LED display modules in numeric and matrix form to display data and simple images. Finally, we put all that together to create a game, a digital thermometer, and more.

Generating Random Numbers

The ability for a program to generate random numbers can be very useful in games and effects. For example, you can use random numbers to play a dice or lottery game with the Arduino, to create lighting effects with LEDs, or to create visual or auditory effects for a quiz game with the Arduino. Unfortunately, the Arduino can't choose a purely random number by itself. You have to help it by providing a *seed,* an arbitrary starting number used in the calculations to generate a random number.

Using Ambient Current to Generate a Random Number

The easiest way to generate a random number with the Arduino is to write a program that reads the voltage from a free (disconnected) analog pin (for example, analog pin zero) with this line in void setup():

```
randomSeed(analogRead(0));
```

Even when nothing is wired to an analog input on the Arduino, static electricity in the environment creates a tiny, measurable voltage. The amount of this voltage is quite random. We can use this measure of ambient voltage as our seed to generate a random number and then allocate it to an integer variable using the random(lower, upper) function. We can use the parameters lower and upper to set the lower and upper limits of the range for the random number. For example, to generate a random number between 100 and 1,000, you would use the following:

```
int a = 0;
a = random(100, 1001);
```

We've used the number 1,001 rather than 1,000 because the 1,001 upper limit is *exclusive*, meaning it's not included in the range.

That said, to generate a random number between 0 and some number, you can just enter the upper limit. Here's how you would generate a random number between 0 and 6:

```
a = random(7);
```

The example sketch in Listing 6-1 would generate random numbers between 0 and 1,000, as well as numbers between 10 and 50:

```
// Listing 6-1
int r = 0;
```

```
void setup()
{
  randomSeed(analogRead(0));
  Serial.begin(9600);
}

void loop()
{
  Serial.print("Random number between zero and 1000 is: ");
  r = random(0, 1001);
  Serial.println(r);
  Serial.print("Random number between ten and fifty is: ");
  r = random(10, 51);
  Serial.println(r);
  delay(1000);
}
```

Listing 6-1: Random number generator

Figure 6-1 shows the result displayed on the Serial Monitor.

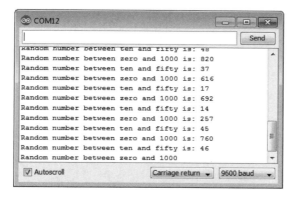

Figure 6-1: Output from Listing 6-1

Now that you know how to generate random numbers, let's put that knowledge to good use by creating an electronic die.

Project #15: Creating an Electronic Die

Our goal is to light one of six LEDs randomly to mimic the throw of a die. We'll choose a random number between 1 and 6, and then turn on the corresponding LED to indicate the result. We'll create a function to select one of six LEDs on the Arduino randomly and to keep the LED on for a certain period of time. When the Arduino running the sketch is turned on or reset, it should rapidly show random LEDs for a specified period of time and then gradually slow until the final LED is lit. The LED matching the resulting randomly chosen number will stay on until the Arduino is reset or turned off.

The Hardware

To build the die, we'll need the following hardware:

- Six LEDs of any color (LED1 to LED6)
- One 560 Ω resistor (R1)
- Various connecting wires
- One medium-sized breadboard
- Arduino and USB cable

The Schematic

Because only one LED will be lit at a time, a single current-limiting resistor can go between the cathodes of the LEDs and GND. Figure 6-2 shows the schematic for our die.

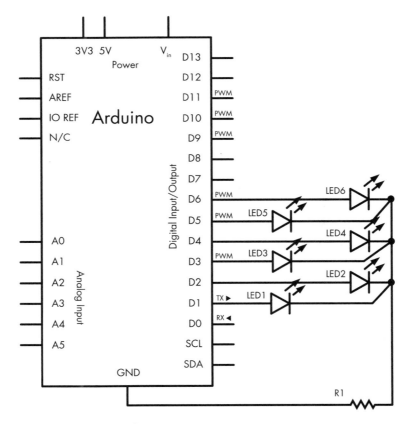

Figure 6-2: Schematic for Project 15

The Sketch

Here's the sketch for our die:

```
// Project 15 - Creating an Electronic Die
void setup()
{
  randomSeed(analogRead(0));        // seed the random number generator
  for ( int z = 1 ; z < 7 ; z++ ) // LEDs on pins 1-6 are output
  {
    pinMode(z, OUTPUT);
  }
}

void randomLED(int del)
{
  int r;
  r = random(1, 7);        // get a random number from 1 to 6
  digitalWrite(r, HIGH); // output to the matching LED on digital pin 1-6
  if (del > 0)
  {
    delay(del);                 // hold the LED on for the delay received
  }
  else if (del == 0)
  {
    do                      // the delay entered was zero, hold the LED on
forever
    {}
    while (1);
  }
  digitalWrite(r, LOW);  // turn off the LED
}

void loop()
{
  int a;
  // cycle the LEDs around for effect
  for ( a = 0 ; a < 100 ; a++ )
  {
    randomLED(50);
  }
  // slow down
  for ( a = 1 ; a <= 10 ; a++ )
  {
    randomLED(a * 100);
  }
  // and stop at the final random number and LED
  randomLED(0);
}
```

❶ `delay(del);`
❷ `else if (del == 0)`
❸ `while (1);`
❹ `for ( a = 1 ; a <= 10 ; a++ )`

Here we use a loop in void setup() to activate the digital output pins. The function randomLED() receives an integer that is used in the delay() function at ❶ to keep the LED turned on for the selected time. If the value of the delay received at ❷ is 0, then the function keeps the LED turned on indefinitely, because we use

```
do {} while (1);
```

at ❸, which loops forever, because 1 is always 1.

To "roll the die," we reset the Arduino to restart the sketch. To create a decreasingly slow change in the LEDs before the final value is displayed, at ❹ we first display a random LED 100 times for 50 milliseconds each time. Next, we slow it down by increasing the delay between LED flashes from 100 to 1,000 milliseconds, with each flash lasting 100 milliseconds. The purpose of this is to simulate the "slowing down" of a die before it finally settles on a value, at which point the Arduino displays the outcome of the roll by keeping one LED lit with this last line:

```
randomLED(0);
```

Modifying the Sketch

We can tinker with this project in many ways. For example, we could add another six LEDs to roll two dice at once, or perhaps display the result using only the built-in LED by blinking it a number of times to indicate the result of the roll. Use your imagination and new skills to have some fun!

A Quick Course in Binary

Most children learn to count using the base-10 system, but computers (and the Arduino) count using the binary number system. *Binary numbers* consist of only 1s and 0s—for example, 10101010. In binary, each digit from right to left represents 2 to the power of the column number in which it appears (which increases from right to left). The products in each column are then added to determine the value of the number.

For example, consider the binary number 11111111, as shown in Table 6-1. To convert the number 11111111 in binary to base 10, we add the totals in each column as listed in the bottom row of the table:

$$128 + 64 + 32 + 16 + 8 + 4 + 2 + 1$$

It is 255. A binary number with eight columns (or *bits*) holds 1 *byte* of data; 1 byte of data can have a numerical value between 0 and 255. The leftmost bit is referred to as the *Most Significant Bit (MSB)*, and the rightmost is the *Least Significant Bit (LSB)*.

Binary numbers are great for storing certain types of data, such as on/off patterns for LEDs, true/false settings, and the statuses of digital outputs. Binary numbers are the building blocks of all types of data in computers.

Table 6-1: Binary to base-10 number conversion example

2^7	2^6	2^5	2^4	2^3	2^2	2^1	2^0	
1	1	1	1	1	1	1	1	Binary
128	64	32	16	8	4	2	1	Base 10

Byte Variables

One way we can store binary numbers is by using a *byte variable*. For example, we can create the byte variable outputs using the following code:

```
byte outputs = B11111111;
```

The B in front of the number tells Arduino to read the number as a binary number (in this case, 11111111) instead of its base-10 equivalent of 255. Listing 6-2 demonstrates this further.

```
// Listing 6-2

byte a;

void setup()
{
  Serial.begin(9600);
}

void loop()
{
  for ( int count = 0 ; count < 256 ; count++ )
  {
    a = count;
    Serial.print("Base-10 = ");
❶   Serial.print(a, DEC);
    Serial.print(" Binary = ");
❷   Serial.println(a, BIN);
    delay(1000);
  }
}
```

Listing 6-2: Binary number demonstration

We display byte variables as base-10 numbers using DEC ❶ or as binary numbers using BIN ❷ as part of the Serial.print() function. After uploading the sketch, you should see output in the Serial Monitor similar to that shown in Figure 6-3.

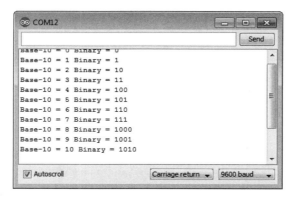

Base-10 = 0 Binary = 0
Base-10 = 1 Binary = 1
Base-10 = 2 Binary = 10
Base-10 = 3 Binary = 11
Base-10 = 4 Binary = 100
Base-10 = 5 Binary = 101
Base-10 = 6 Binary = 110
Base-10 = 7 Binary = 111
Base-10 = 8 Binary = 1000
Base-10 = 9 Binary = 1001
Base-10 = 10 Binary = 1010

Figure 6-3: Output from Listing 6-2

Increasing Digital Outputs with Shift Registers

The Arduino board has 13 digital pins that we can use as outputs—but sometimes 13 just isn't enough. To add outputs, we can use a *shift register* and still have plenty of room left on the Arduino for outputs. A shift register is an integrated circuit (IC) with eight digital output pins that can be controlled by sending a byte of data to the IC. For our projects, we will be using the 74HC595 shift register shown in Figure 6-4.

Figure 6-4: The 74HC595 shift register IC

The 74HC595 shift register has eight digital outputs that can operate like the Arduino digital output pins. The shift register itself takes up three Arduino digital output pins, so the net gain is five output pins.

The principle behind the shift register is simple: We send 1 byte of data (8 bits) to the shift register, and it turns on or off the matching eight outputs based on the 1 byte of data. The bits representing the byte of data match the output pins in order from highest to lowest. So the leftmost bit of the data represents output pin 7 of the shift register, and the rightmost bit of the data represents output pin 0. For example, if we send B10000110 to the shift register, then it will turn on outputs 7, 2, and 1 and will turn off outputs 0 and 3–6 until the next byte of data is received or the power is turned off.

More than one shift register can also be connected together to provide an extra eight digital output pins for every shift register attached to the same three Arduino pins; this makes shift registers very convenient when you want to control lots of LEDs. Let's do that now by creating a binary number display.

Project #16: Creating an LED Binary Number Display

In this project, we'll use eight LEDs to display binary numbers from 0 to 255. Our sketch will use a for loop to count from 0 to 255 and will send each value to the shift register, which will use LEDs to display the binary equivalent of each number.

The Hardware

The following hardware is required:

- One 74HC595 shift register IC
- Eight LEDs (LED1 to LED8)
- Eight 560 Ω resistors (R1 to R8)
- One breadboard
- Various connecting wires
- Arduino and USB cable

Connecting the 74HC595

Figure 6-5 shows the schematic symbol for the 74HC595.

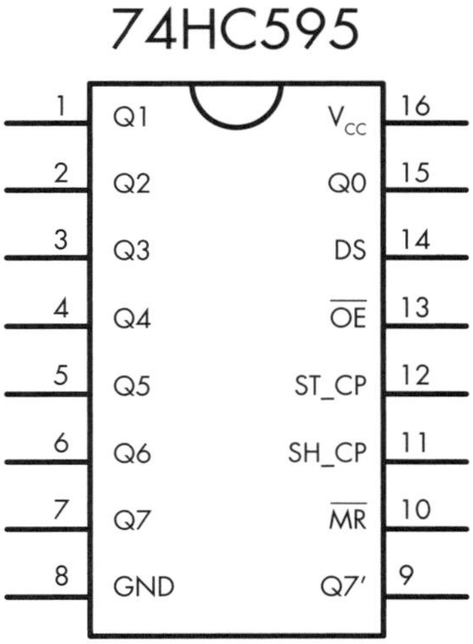

Figure 6-5: 74HC595 schematic symbol

There are 16 pins on our shift register:

- Pins 15 and 1 to 7 are the eight output pins that we control (labeled *Q0* to *Q7*, respectively).
- Q7 outputs the first bit sent to the shift register, down to Q0, which outputs the last.

- Pin 8 connects to GND.
- Pin 9 is "data out" and is used to send data to another shift register if one is present.
- Pin 10 is always connected to 5 V (for example, the 5 V connector on the Arduino).
- Pins 11 and 12 are called *clock* and *latch*.
- Pin 13 is called *output enable* and is usually connected to GND.
- Pin 14 is for incoming bit data sent from the Arduino.
- Pin 16 is used for power: 5 V from the Arduino.

To give you a sense of the way pins are oriented, the semicircular notch on the left end of the body of the shift register IC shown in Figure 6-4 lies between pins 1 and 16.

The pins are numbered sequentially around the body in a counter-clockwise direction, as shown in Figure 6-6, the schematic for our LED binary number display.

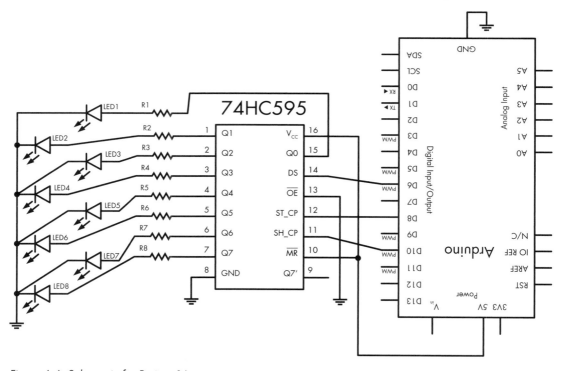

Figure 6-6: Schematic for Project 16

NOTE   *Once you have finished with this example circuit, keep it assembled. We'll use it again with the forthcoming project.*

The Sketch

And now for the sketch:

```
// Project 16 - Creating an LED Binary Number Display
#define DATA  6          // digital 6 to pin 14 on the 74HC595
#define LATCH 8          // digital 8 to pin 12 on the 74HC595
#define CLOCK 10         // digital 10 to pin 11 on the 74HC595

void setup()
{
   pinMode(LATCH, OUTPUT);
   pinMode(CLOCK, OUTPUT);
   pinMode(DATA, OUTPUT);
}

void loop()
{
   int i;
   for ( i = 0; i < 256; i++ )
   {
     digitalWrite(LATCH, LOW);
     shiftOut(DATA, CLOCK, MSBFIRST, i);
     digitalWrite(LATCH, HIGH);
     delay(200);
   }
}
```

In this sketch, we set the three pins connected to the shift register as outputs in void setup() and then add a loop in void loop() that counts from 0 to 255 and repeats. The magic lies inside the loop. When we send a byte of data (for example, 240, or B11110000) to the shift register in the for loop, three things happen:

- The latch pin 12 is set to LOW (that is, a low signal is applied to it from the Arduino digital output pin 8). This is preparation for setting output pin 12 to HIGH, which latches the data to the output pins after shiftOut has completed its task.

- We send the byte of data (for example, B11110000) from Arduino digital pin 6 to the shift register and tell the shiftOut function from which direction to interpret the byte of data. For example, if you selected LSBFIRST, then LEDs 1 to 4 would turn on and the others would turn off. If you used MSBFIRST, then LEDs 5 to 8 would turn on and the others would turn off.

- Finally, the latch pin 12 is set to HIGH (5 V is applied to it). This tells the shift register that all the bits are shifted in and ready. At this point it alters its output to match the data received.

In this project we'll use random numbers, the Serial Monitor, and the circuit created in Project 16 to create a binary quiz game. The Arduino will display a random binary number using the LEDs, and then you will enter the decimal version of the binary number using the Serial Monitor. The Serial Monitor will then tell you whether your answer is correct and the game will continue with a new number.

The Algorithm

The algorithm can be divided into three functions. The displayNumber() function will display a binary number using the LEDs. The getAnswer() function will receive a number from the Serial Monitor and display it to the user. Finally, the checkAnswer() function will compare the user's number to the random number generated and display the correct/incorrect status and the correct answer if the guess was incorrect.

The Sketch

The sketch generates a random number between 0 and 255, displays it in binary using the LEDs, asks the user for his or her answer, and then displays the results in the Serial Monitor. You've already seen all the functions used in the sketch, so although there's a lot of code here, it should look familiar. We'll dissect it with comments within the sketch and some commentary following.

```
// Project 17 - Making a Binary Quiz Game

#define DATA   6              // connect to pin 14 on the 74HC595
#define LATCH  8              // connect to pin 12 on the 74HC595
#define CLOCK 10              // connect to pin 11 on the 74HC595

int number = 0;
int answer = 0;

❶ void setup()
  {
    pinMode(LATCH, OUTPUT);    // set up the 74HC595 pins
    pinMode(CLOCK, OUTPUT);
    pinMode(DATA, OUTPUT);
    Serial.begin(9600);
    randomSeed(analogRead(0)); // initialize the random number generator
    displayNumber(0);          // clear the LEDs
  }

❷ void displayNumber(byte a)
  {
    // sends byte to be displayed on the LEDs
    digitalWrite(LATCH, LOW);
```

```
      shiftOut(DATA, CLOCK, MSBFIRST, a);
      digitalWrite(LATCH, HIGH);
   }

❸ void getAnswer()
   {
      // receive the answer from the player
      int z = 0;
      Serial.flush();
      while (Serial.available() == 0)
      {
         // do nothing until something comes into the serial buffer
      }
      // one character of serial data is available, begin calculating
      while (Serial.available() > 0)
      {
         // move any previous digit to the next column on the left; in
         // other words, 1 becomes 10 while there is data in the buffer
         answer = answer * 10;
         // read the next number in the buffer and subtract the character '0'
         // from it to convert it to the actual integer number
         z = Serial.read() - '0';
         // add this digit into the accumulating value
         answer = answer + z;
         // allow a short delay for any more numbers to come into Serial.available
         delay(5);
      }
      Serial.print("You entered: ");
      Serial.println(answer);
   }

❹ void checkAnswer()
   {
      //check the answer from the player and show the results
      if (answer == number)     // Correct!
      {
        Serial.print("Correct! ");
        Serial.print(answer, BIN);
        Serial.print(" equals ");
        Serial.println(number);
        Serial.println();
      }
      else                      // Incorrect
      {
        Serial.print("Incorrect, ");
        Serial.print(number, BIN);
        Serial.print(" equals ");
        Serial.println(number);
        Serial.println();
      }
      answer = 0;
      delay(10000); // give the player time to review his or her answers
   }
```

```
void loop()
{
  number = random(256);
  displayNumber(number);
  Serial.println("What is the binary number in base-10? ");
  getAnswer();
  checkAnswer();
}
```

Let's review how the sketch works. At ❶, void setup() configures the digital output pins to use the shift register, starts the Serial Monitor, and seeds the random number generator. At ❷, the custom function displayNumber() accepts a byte of data and sends it to the shift register, which uses LEDs to display the byte in binary form via the attached LEDs (as in Project 16). At ❸, the custom function getAnswer() accepts a number from the user via the Serial Monitor (as in Project 14) and displays it, as shown in Figure 6-7.

The function checkAnswer() at ❹ compares the number entered by the player in getAnswer() against the random number generated by the sketch in void loop(). The player is then advised of a correct or incorrect answer with corresponding binary and decimal values. Finally, in the main void loop() from which the program runs, the Arduino generates the random binary number for the quiz, then calls the matching functions to display it with hardware, and then receives and checks the player's answer.

Figure 6-7 shows the game in play in the Serial Monitor.

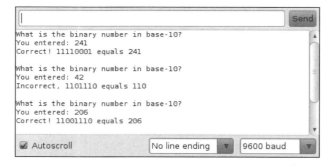

Figure 6-7: Project 17 in play

Arrays

An *array* is a set of variables or values grouped together so that they can be referenced as a whole. When dealing with lots of related data, you'll find it a good idea to use arrays to keep your data organized.

Defining an Array

Each item in an array is called an *element*. For example, suppose six float variables contain temperatures taken over the last six hours; instead of giving them all separate names, we can define an array called temperatures with six elements like this:

```
float temperatures[6];
```

We can also insert values when defining the array. When we do that, we don't need to define the array size. Here's an example:

```
float temperatures[]={11.1, 12.2, 13.3, 14.4, 15.5, 16.6};
```

Notice that this time we didn't explicitly define the size of the array within the square brackets ([]); instead, its size is deduced based on the number of elements set by the values inside the curly brackets ({}).

Referring to Values in an Array

We count the elements in an array beginning from the left and starting from 0; the temperatures[] array has elements numbered 0 to 5. We can refer to individual values within an array by inserting the number of the element in the square brackets. For example, to change the first element in temperatures[] (currently 16.6) to 12.34, we would use this:

```
temperatures[0] = 12.34;
```

Writing to and Reading from Arrays

In Listing 6-3, we demonstrate writing values to and reading values from an array of five elements. The first for loop in the sketch writes a random number into each of the array's elements, and the second for loop retrieves the elements and displays them in the Serial Monitor.

```
// Listing 6-3

void setup()
{
  Serial.begin(9600);
  randomSeed(analogRead(0));
}
int array[5];     // define our array of five integer elements
void loop()
{
  int i;
  Serial.println();
  for ( i = 0 ; i < 5 ; i++ )   // write to the array
  {
    array[i] = random(10);      // random numbers from 0 to 9
  }
  for ( i = 0 ; i < 5 ; i++ )   // display the contents of the array
  {
    Serial.print("array[");
```

```
    Serial.print(i);
    Serial.print("] contains ");
    Serial.println(array[i]);
  }
  delay(5000);
}
```

Listing 6-3: Array read/write demonstration

Figure 6-8 shows the output of this sketch in the Serial Monitor.

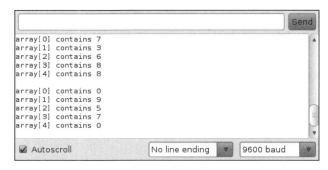

Figure 6-8: Listing 6-3 in action

Now that you know how to work with binary numbers, shift registers, and arrays, it's time you put that knowledge to work. In the next section, we'll wire up some digital number displays.

Seven-Segment LED Displays

LEDs are fun, but there are limits to the kinds of data that can be displayed with individual lights. In this section we'll begin working with numeric digits in the form of seven-segment LED displays, as shown in Figure 6-9.

Figure 6-9: Seven-segment display modules

These displays are perfect for displaying numbers, and that's why you'll find them used in digital alarm clocks, speedometers, and other numeric displays. Each module in a seven-segment LED display consists of eight LEDs. The modules are also available in different colors. To reduce the number of pins used by the display, all of the anodes or cathodes of the LEDs are connected together and are called *common-anode* or *common-cathode,* respectively. Our projects will use common-cathode modules.

The display's LEDs are labeled *A* to *G* and *DP* (for the decimal point). There is an anode pin for each LED segment, and the cathodes are connected to one common cathode pin. The layout of seven-segment LED displays is always described as shown in Figure 6-10, with LED segment A at the top, B to its right, and so on. So, for example, if you wanted to display the number 7, then you would apply current to segments A, B, and C.

The pins on each LED display module can vary, depending on the manufacturer, but they always follow the basic pattern shown in Figure 6-10. When you use one of these modules, always get the data sheet for the module from the retailer to help save you time determining which pins are which.

We'll use the schematic symbol shown in Figure 6-11 for our seven-segment LED display modules.

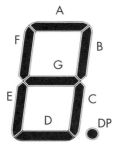

Figure 6-10: LED map for a typical seven-segment display module

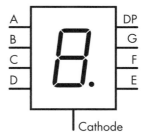

Figure 6-11: Schematic symbol for a seven-segment display module

Controlling the LED

We'll control the LED display using the method discussed in Project 17, by connecting pins A through DP to the shift register outputs Q0 to Q7. Use the matrix shown in Table 6-2 as a guide to help you determine which segments to turn on and off to display a particular number or letter.

The top row in the matrix is the shift register output pin that controls the segments on the second row. Each row below this shows the digit that can be displayed with the corresponding binary and decimal value to send to the shift register.

Table 6-2: Display Segment Matrix

SR	Q0	Q1	Q2	Q3	Q4	Q5	Q6	Q7	
Segment	A	B	C	D	E	F	G	DP	Decimal
0	1	1	1	1	1	1	0	0	252
1	0	1	1	0	0	0	0	0	96
2	1	1	0	1	1	0	1	0	218
3	1	1	1	1	0	0	1	0	242
4	0	1	1	0	0	1	1	0	102
5	1	0	1	1	0	1	1	0	182
6	1	0	1	1	1	1	1	0	190
7	1	1	1	0	0	0	0	0	224
8	1	1	1	1	1	1	1	0	254
9	1	1	1	1	0	1	1	0	246
A	1	1	1	0	1	1	1	0	238
B	0	0	1	1	1	1	1	0	62
C	1	0	0	1	1	1	0	0	156
D	0	1	1	1	1	0	1	0	122
E	1	0	0	1	1	1	1	0	158
F	1	0	0	0	1	1	1	0	142

For example, to display the digit 7 as shown in Figure 6-12, we need to turn on LED segments A, B, and C, which correspond to the shift register outputs Q0, Q1, and Q2. Therefore, we will send the byte B1110000 into the shift register (with shiftOut set to LSBFIRST) to turn on the first three outputs that match the desired LEDs on the module.

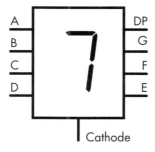

Figure 6-12: Displaying the digit 7

In the next example, we'll create a circuit that displays, in turn, the digits 0 through 9 and then the letters A through F. The cycle repeats with the decimal-point LED turned on.

In this project we'll assemble a circuit to use a single-digit display.

The Hardware

The following hardware is required:

- One 74HC595 shift register IC
- One common-cathode seven-segment LED display
- Eight 560 Ω resistors (R1 to R8)
- One large breadboard
- Various connecting wires
- Arduino and USB cable

The Schematic

The schematic is shown in Figure 6-13.

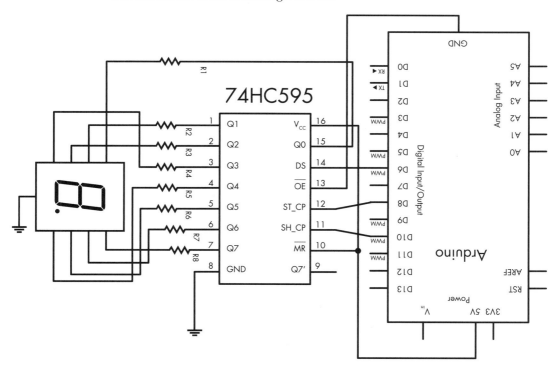

Figure 6-13: Schematic for Project 18

When wiring the LED module to the shift register, LED pins A through G connect to pins Q0 through Q6, respectively, and DP connects to Q7.

The Sketch

In the sketch for Project 18, we store the decimal values (see Table 6-2) in the int digits[] array. In the void loop, we send these values to the shift register in sequential order at ❶ and then repeat the process with the decimal point on by adding 1 to the value sent to the shift register at ❷:

```
// Project 18 - Creating a Single-Digit Display
#define DATA  6                    // connect to pin 14 on the 74HC595
#define LATCH 8                    // connect to pin 12 on the 74HC595
#define CLOCK 10                   // connect to pin 11 on the 74HC595

// set up the array with the segments for 0 to 9, A to F (from Table 6-2)
int digits[] = {252, 96, 218, 242, 102, 182, 190, 224, 254, 246, 238, 62, 156,
122, 158, 142};

void setup()
{
  pinMode(LATCH, OUTPUT);
  pinMode(CLOCK, OUTPUT);
  pinMode(DATA, OUTPUT);
}

void loop()
{
  int i;
  for ( i = 0 ; i < 16 ; i++ )   // display digits 0-9, A-F
  {
    digitalWrite(LATCH, LOW);
❶ shiftOut(DATA, CLOCK, LSBFIRST, digits[i]);
    digitalWrite(LATCH, HIGH);
    delay(250);
  }
  for ( i = 0 ; i < 16 ; i++ )   // display digits 0-9, A-F with DP
  {
    digitalWrite(LATCH, LOW);
❷ shiftOut(DATA, CLOCK, LSBFIRST, digits[i]+1); // +1 is to turn on the DP bit
    digitalWrite(LATCH, HIGH);
    delay(250);
  }
}
```

Seven-segment LCD displays are bright and easy to read. For example, Figure 6-14 shows the digit 9 with the decimal point displayed as a result of this sketch.

Figure 6-14: Digit displayed by Project 18

Displaying Double Digits

To use more than one shift register to control additional digital outputs, connect pin 9 of the 74HC595 (which receives data from the Arduino) to pin 14 of the second shift register. Once you've made this connection, 2 bytes of data will sent: the first to control the second shift register and the second to control the first shift register. Here's an example:

```
digitalWrite(LATCH, LOW);
shiftOut(DATA, CLOCK, MSBFIRST, 254); // data for second 74HC595
shiftOut(DATA, CLOCK, MSBFIRST, 254); // data for first 74HC595
digitalWrite(LATCH, HIGH);
```

Project #19: Controlling Two Seven-Segment LED Display Modules

This project will show you how to control two, seven-segment LED display modules so that you can display two-digit numbers.

The Hardware

The following hardware is required:

- Two 74HC595 shift register ICs
- Two common-cathode seven-segment LED displays
- Sixteen 560 Ω resistors (R1 to 16)
- One large breadboard
- Various connecting wires
- Arduino and USB cable

The Schematic

Figure 6-15 shows the schematic for two display modules.

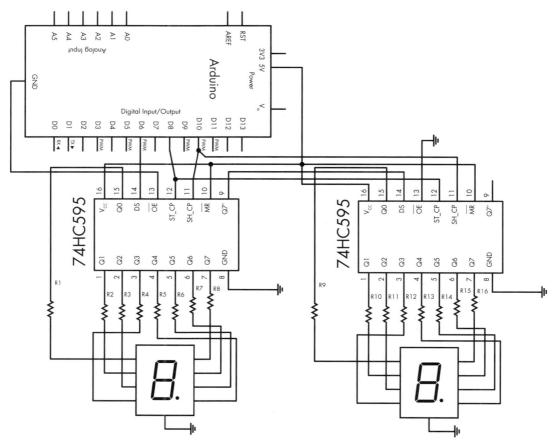

Figure 6-15: Schematic for Project 19

Note that the shift registers' data and clock pins are connected to each other and then to the Arduino. The data line from Arduino digital pin 6 runs to shift register 1, and then a link from pin 9 of shift register 1 runs to pin 14 of shift register 2.

To display a number between 0 and 99, we'll need a more complicated sketch. If a number is less than 10, then we can just send the number followed by a 0, as the right digit displays the number and the left digit displays 0. However, if the number is greater than 10, then we need to determine each of the number's two digits and send each to the shift registers separately. To make this process easier, we'll use the math function modulo.

Modulo

Modulo is a function that returns the remainder of a division operation. For example, 10 modulo (or *mod*) 7 equals 3—in other words, the remainder of 10 divided by 7 equals 3. We use the percent sign (%) to represent modulo. The following example uses modulo in a sketch:

```
int a = 8;
int b = 3;
c = a % b;
```

In this example, the value of c will be 2. So to determine a two-digit number's right-hand digit, we use the modulo function, which returns the remainder when dividing the two numbers.

To automate displaying a single- or double-digit number, we'll create the function displayNumber() for our sketch. We use modulo as part of this function to separate the digits of a two-digit number. For example, to display the number *23*, we first isolate the left-hand digit by dividing 23 by 10, which equals 2 (and a fraction that we can ignore). To isolate the right-hand digit, we perform 23 modulo 10, which equals 3.

```
// Project 19 - Controlling Two Seven-Segment LED Display Modules
// set up the array with the segments for 0 to 9, A to F (from Table 6-2)

#define DATA  6          // connect to pin 14 on the 74HC595
#define LATCH 8          // connect to pin 12 on the 74HC595
#define CLOCK 10         // connect to pin 11 on the 74HC595

void setup()
{
  pinMode(LATCH, OUTPUT);
  pinMode(CLOCK, OUTPUT);
  pinMode(DATA, OUTPUT);
}

int digits[] = {252, 96, 218, 242, 102, 182, 190, 224, 254, 246, 238, 62, 156,
122, 158, 142};
void displayNumber(int n)
{
  int left, right=0;
  if (n < 10)
  {
    digitalWrite(LATCH, LOW);
    shiftOut(DATA, CLOCK, LSBFIRST, digits[n]);
    shiftOut(DATA, CLOCK, LSBFIRST, 0);
    digitalWrite(LATCH, HIGH);
  }
  else if (n >= 10)
  {
    right = n % 10; // remainder of dividing the number to display by 10
    left = n / 10;  // quotient of dividing the number to display by 10
    digitalWrite(LATCH, LOW);
```

❶ `if (n < 10)`

❷ `right = n % 10;`

```
        shiftOut(DATA, CLOCK, LSBFIRST, digits[right]);
        shiftOut(DATA, CLOCK, LSBFIRST, digits[left]);
        digitalWrite(LATCH, HIGH);
      }
    }
❸ void loop()
    {
      int i;
      for ( i = 0 ; i < 100 ; i++ )
      {
        displayNumber(i);
        delay(100);
      }
    }
```

At ❶, the function checks to see if the number to be displayed is less than 10. If so, it sends the data for the number and a blank digit to the shift registers. However, if the number is greater than 10, then the function uses modulo and division at ❷ to separate the digits and then sends them to the shift registers separately. Finally, in void loop() at ❸ we set up and call the function to display the numbers from 0 to 99.

Project #20: Creating a Digital Thermometer

In this project we'll add the TMP36 temperature sensor we created in Chapter 4 to the double-digit circuit constructed for Project 19 to create a digital thermometer. The algorithm is simple: we read the voltage returned from the TMP36 (using the method from Project 12) and convert the reading to degrees Celsius.

The Hardware

The following hardware is required:

- The double-digit circuit from Project 19
- One TMP36 temperature sensor

Connect the center output lead of the TMP36 to analog pin 5, the left lead to 5 V, and the right lead to GND, and you're ready to measure.

The Sketch

Here is the sketch:

```
// Project 20 - Creating a Digital Thermometer
#define DATA  6          // connect to pin 14 on the 74HC595
#define LATCH 8          // connect to pin 12 on the 74HC595
#define CLOCK 10         // connect to pin 11 on the 74HC595
```

```
int temp = 0;
float voltage = 0;
float celsius = 0;
float sensor = 0;
int digits[]={
  252, 96, 218, 242, 102, 182, 190, 224, 254, 246, 238, 62, 156, 122, 158, 142};

void setup()
{
  pinMode(LATCH, OUTPUT);
  pinMode(CLOCK, OUTPUT);
  pinMode(DATA, OUTPUT);
}

void displayNumber(int n)
{
  int left, right = 0;
  if (n < 10)
  {
    digitalWrite(LATCH, LOW);
    shiftOut(DATA, CLOCK, LSBFIRST, digits[n]);
    shiftOut(DATA, CLOCK, LSBFIRST, digits[0]);
    digitalWrite(LATCH, HIGH);
  }
  if (n >= 10)
  {
    right = n % 10;
    left = n / 10;
    digitalWrite(LATCH, LOW);
    shiftOut(DATA, CLOCK, LSBFIRST, digits[right]);
    shiftOut(DATA, CLOCK, LSBFIRST, digits[left]);
    digitalWrite(LATCH, HIGH);

  }
}

void loop()
{
  sensor = analogRead(5);
  voltage = (sensor * 5000) / 1024; // convert raw sensor value to millivolts
  voltage = voltage - 500;         // remove voltage offset
  celsius = voltage / 10;          // convert millivolts to Celsius
  temp = int(celsius); // change the floating point temperature to an int
  displayNumber(temp);
  delay(500);
}
```

The sketch is simple and borrows code from previous projects: displayNumber() from Project 19 and the temperature calculations from Project 12. The delay(500); function in the second-to-last line of the sketch keeps the display from changing too quickly when the temperature fluctuates.

LED Matrix Display Modules

If you enjoyed experimenting with blinking LEDs, then you're going to love LED matrix modules. An *LED matrix module* consists of several rows and columns of LEDs that you can control individually or in groups. The example that we'll use (shown in Figure 6-16) has eight rows and eight columns of red LEDs for a total of 64 LEDs. This LED matrix is a common-cathode display.

In this project we'll build one circuit, and then we'll use various sketches to create different effects.

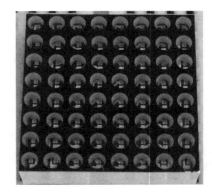

Figure 6-16: LED matrix

The LED Matrix Schematic

The schematic symbol for the matrix looks a bit complex, as shown in Figure 6-17.

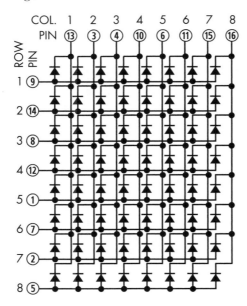

Figure 6-17: LED matrix schematic symbol

Notice that the numbering of the rows and columns of LEDs does not match the order of the pins beneath the matrix; see the circled pin numbers in Figure 6-17. On the underside of our matrix, pin 1 is indicated at the bottom right. For example, in Figure 6-18, you can see a tiny *1* printed below the pin.

Figure 6-18: Pins on the LED matrix with a
tiny 1 printed below the bottom-right pin

The pins on the LED matrix are numbered clockwise, with pin 8 at
the bottom left and pin 16 at the top right. We'll control the matrix with
two 74HC595 shift registers in a way that's similar to the method used in
Project 19.

Figure 6-19 shows the schematic for our project. Resistors R1 through
R8 are 560 ohms each. One shift register controls the rows of LEDs, and
the other controls the columns. The LED matrix (not shown here) is connected to the outputs at the bottom of this schematic with the pin connections listed in Table 6-3.

Table 6-3: Matrix to 74HC595 link table

Row SR Pin	Matrix Pin	Column SR Pin	Matrix Pin
15	9	15	13
1	14	1	3
2	8	2	4
3	12	3	10
4	1	4	6
5	7	5	11
6	2	6	15
7	5	7	16

Making the Connections

Make the connections between the shift registers and the LED matrix (don't
forget the resistors between the output pins of the shift register, which controls the matrix rows, and the matrix row pins), as shown in Table 6-3.

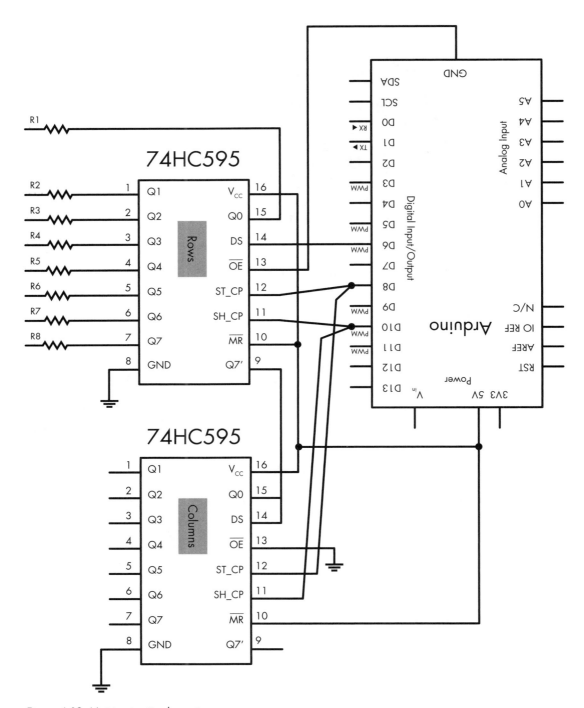

Figure 6-19: Matrix circuit schematic

Now to get the matrix working. The shift register marked "rows" in Figure 6-19 allows current to flow into each row of the matrix, and the "columns" shift register allows current to flow from each of the columns of the matrix to GND. We'll use a simple sketch to test the setup in the next project. To turn on an LED, we need to control the matching row and column pins of the shift registers.

Bitwise Arithmetic

We can use *bitwise arithmetic* to manipulate integer and byte variables using their binary representations. The value in using bitwise arithmetic, rather than base-10 arithmetic, is that bitwise can help increase the speed of controlling digital input and output pins and can compare numbers in binary form.

There are six *bitwise operators*: AND, OR, XOR, NOT, bitshift left, and bitshift right. Each is discussed in the following sections. (Although these examples use binary numbers, the same operations could be performed using integers and bytes.)

The Bitwise AND Operator

The AND operator (&) is used to compare two binary numbers bit by bit. If the bits in the same column of both numbers are 1, then the resulting value's bit is set to 1; if the bits are not 1, then the result is set to 0. Consider, for example, these 3-byte variables:

```
byte a = B00110011;
byte b = B01010001;
byte c = 0;
```

The result of the comparison

```
c = a & b;
```

will be 00010001. Expanding this result in a text-comment diagram can show this in more detail:

```
byte a = B00110011;
//         ||||||||
byte b = B01010001;
//         ||||||||
//         00010001  c = a & b; // c is equal to a 'AND'ed with b
```

The Bitwise OR Operator

The OR operator (|) compares two binary numbers, but instead of returning a 1 if both numbers in a column are 1, it returns a 1 if *either* value in the column is 1. If both numbers in a column are 0, then 0 is returned.

Using the same demonstration bytes as before,

```
byte a = B00110011;
byte b = B01010001;
byte c = 0;
```

the result of the comparison

```
c = a | b;
```

will be 01110011.

The Bitwise XOR Operator

The XOR operator (^) returns a 1 result if the bits are different and a 0 result if they are the same.

Using our demonstration bytes again,

```
byte a = B00111100;
byte b = B01011010;
byte c = 0;
```

the result of the comparison

```
c = a ^ b;
```

will be 01100110.

The Bitwise NOT Operator

The NOT operator (~) simply reverses, or flips, the bits in each column: 0s are changed to 1s, and 1s are changed to 0s. Consider this example: If we store a bitwise NOT of byte a in byte b like so,

```
byte a = B00111100;
byte b = 0;
b = ~a;
```

then b contains 11000011.

Bitshift Left and Right

The bitshift left («) and bitshift right (») operators move bits to the left or right by a certain number of positions. For example, if the contents of a are shifted left four spaces, like so,

```
byte a = B00100101;
byte b = a << 4;
```

then the result is the value 01010000 for b. The bits in a are moved left four spaces, and the empty spaces are filled with 0s.

If we shift in the other direction, like so,

```
byte a = B11110001;
byte b = a >> 4;
```

then the value for b will be 00001111.

Project #21: Creating an LED Matrix

The purpose of this project is to demonstrate the use of the LED matrix; we'll turn on every second column and row in the matrix, as shown in Figure 6-20.

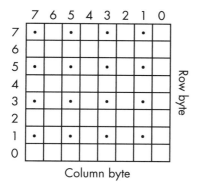

Figure 6-20: Checkerboard matrix template

To create this display pattern, we'll send B10101010 to the rows shift register and ~B10101010 to the columns shift register. The 1s and 0s in each byte match the rows and columns of the matrix.

NOTE *Note the use of the bitwise NOT (~) on the columns byte. A columns shift register bit needs to be 0 to turn on an LED from the column connections. However, a rows shift register bit needs to be 1 to turn on the LED from the rows connections. Therefore, we use the bitwise arithmetic ~ to invert the byte of data being sent to the second shift register that drives the columns.*

To create the effect shown in Figure 6-20, use the following sketch:

```
// Project 21 - Creating an LED Matrix
#define DATA  6      // connect to pin 14 on the 74HC595
#define LATCH 8      // connect to pin 12 on the 74HC595
#define CLOCK 10     // connect to pin 11 on the 74HC595
```

```
void setup()
{
  pinMode(LATCH, OUTPUT);
  pinMode(CLOCK, OUTPUT);
  pinMode(DATA, OUTPUT);
}

void loop()
{
  digitalWrite(LATCH, LOW);
  shiftOut(DATA, CLOCK, MSBFIRST, ~B10101010); // columns
  shiftOut(DATA, CLOCK, MSBFIRST,  B10101010); // rows
  digitalWrite(LATCH, HIGH);
  do {} while (1); // do nothing
}
```

The result is shown in Figure 6-21. We have turned on every other LED inside the matrix to form a checkerboard pattern.

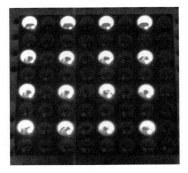

Figure 6-21: Result of Project 21

Project #22: Creating Images on an LED Matrix

To display an image or pattern on the LED matrix, we need a function that turns on only one LED at a time. However, to display an image, we need to turn on and off the LEDs that represent the image very quickly in order to create *persistence of vision (POV)* effects. Persistence of vision produces an image that remains in our eye for a fraction of a second after the image is gone. We can harness this effect to create custom images by "scanning" the matrix to display one row of LEDs at a time, very quickly. This technique can be useful for creating animation, for displaying data, and for creating various other artistic effects.

We will demonstrate controlling individual LEDs in the next two projects. In the following sketch, the function void setLED() accepts a row and column number and the duration to keep the LED turned on. Then it randomly turns on LEDs one at a time.

```
// Project 22 - Creating Images on an LED Matrix
#define DATA  6        // connect to pin 14 on the 74HC595
#define LATCH 8        // connect to pin 12 on the 74HC595
#define CLOCK 10       // connect to pin 11 on the 74HC595

void setup()
{
  pinMode(LATCH, OUTPUT);
  pinMode(CLOCK, OUTPUT);
  pinMode(DATA, OUTPUT);
  randomSeed(analogRead(0));
}

int binary[] = {1, 2, 4, 8, 16, 32, 64, 128};
int r, c = 0;

void setLED(int row, int column, int del)
{
  digitalWrite(LATCH, LOW);
  shiftOut(DATA, CLOCK, MSBFIRST, ~binary[column]);    // columns
  shiftOut(DATA, CLOCK, MSBFIRST,  binary[row]);       // rows
  digitalWrite(LATCH, HIGH);
  delay(del);
}

void loop()
{
  r = random(8);
  c = random(8);
  setLED(r, c, 1);  // set a random row and column on for 1 millisecond
}
```

Instead of sending binary numbers directly to the shiftOut() functions to control which lights are turned on, we use a lookup table in the form of an array, int binary[], at ❶. This lookup table contains the decimal equivalent for each bit of the byte sent to the shift register. For example, to turn on the LED at row 4, column 4, we send binary[3] (which is 8, or B00001000) to both shift registers (with the addition of ~ for the column). This is a convenient way of converting desired row or column numbers into a form the shift register can use; you don't need to think in binary, just the row or column you need to turn on.

By running Project 22 with a delay of 1 millisecond, the LEDs turn on and off so fast that the eye perceives that more than one LED is lit at a time; this demonstrates the concept of POV: It creates the illusion that more than one LED is lit at once, when in fact only one LED is lit at once. An example of this is shown in Figure 6-22.

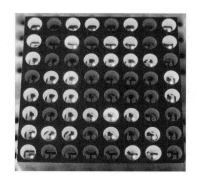

Figure 6-22: Project 22 at work

In this project we'll display the image on the matrix shown in Figure 6-23.

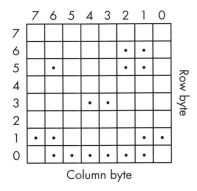

Figure 6-23: Layout for our POV display example

We can define each row as a binary number and place these numbers in an array:

```
byte smile[] = {B00000000,
                B00000110,
                B01000110,
                B00000000,
                B00011000,
                B00000000,
                B11000011,
                B01111110};
```

Notice how the 1s in the array resemble the lit LEDs in the layout in Figure 6-23. Our use of binary numbers makes developing an image very easy. By experimenting with MSBFIRST and LSBFIRST in the shiftOut() functions, you can flip each row around, as shown next. We use a for loop at ❶ to display each row in turn:

```
// Project 23 - Displaying an Image on an LED Matrix
#define DATA  6    // connect to pin 14 on the 74HC595
#define LATCH 8    // connect to pin 12 on the 74HC595
#define CLOCK 10   // connect to pin 11 on the 74HC595

byte smile[] = {B00000000, B00000110, B01000110, B00000000, B00011000,
B00000000, B11000011, B01111110};
int binary[] = {1, 2, 4, 8, 16, 32, 64, 128};

void setup()
{
  pinMode(LATCH, OUTPUT);
  pinMode(CLOCK, OUTPUT);
```

```
    pinMode(DATA, OUTPUT);
}

void loop()
{
  int i;
❶ for ( i = 0 ; i < 8 ; i++ )
  {
    digitalWrite(LATCH, LOW);
    shiftOut(DATA, CLOCK, MSBFIRST, ~smile[i]);  // columns
    shiftOut(DATA, CLOCK, LSBFIRST,  binary[i]); // rows
    digitalWrite(LATCH, HIGH);
    delay(1);
  }
}
```

The result is a winking smiley face, as shown in Figure 6-24.

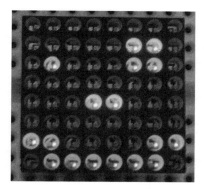

Figure 6-24: Result of Project 23

Project #24: Animating an LED Matrix

Using some bitwise arithmetic, we can scroll the image we created in
Project 23 off the display in either direction. For example, to have the face
move to the right until its gone, we display the image, then shift each bit in
the row over by one using >>, and then display the image again.

The Sketch

Here's the sketch for our animation demonstration:

```
// Project 24 - Animating an LED Matrix
#define DATA 6    // connect to pin 14 on the 74HC595
#define LATCH 8   // connect to pin 12 on the 74HC595
#define CLOCK 10  // connect to pin 11 on the 74HC595
```

```
byte smile[] = {B00000000, B00000110, B01000110, B00000000, B00011000,
B00000000, B11000011, B01111110};
int binary[] = {1, 2, 4, 8, 16, 32, 64, 128};

void setup()
{
  pinMode(LATCH, OUTPUT);
  pinMode(CLOCK, OUTPUT);
  pinMode(DATA, OUTPUT);
}

void loop()
{
  int a, hold, shift;
❶  for ( shift = 0 ; shift < 9 ; shift++ )
  {
❷    for ( hold = 0 ; hold < 25 ; hold++ )
    {
      for ( a = 0 ; a < 8 ; a++ )
      {
        digitalWrite(LATCH, LOW);
❸        shiftOut(DATA, CLOCK, MSBFIRST, ~smile[a]>>shift); // columns
        shiftOut(DATA, CLOCK, LSBFIRST,  binary[a]);        // rows
        digitalWrite(LATCH, HIGH);
        delay(1);
      }
    }
  }
}
```

The sketch holds the image on the matrix for 25 display cycles using a
for loop at ❷. The variable shift is the amount by which each byte will shift
to the right. After each loop has completed, the variable shift is increased
by 1 as shown at ❶. Next, the display cycles repeat, and the image moves to
the right by one LED column. By changing MSBFIRST to LSBFIRST in the third
for loop at ❸, we can change the direction that the face scrolls.

Looking Ahead

In this chapter you have learned a lot of fundamental skills that will be
used over and over in your own projects. LED displays are relatively hardy,
so enjoy experimenting with them and making various display effects.
However, there is a limit to what can be displayed, so in the next chapter
we make use of much more detailed display methods for text and graphics.

7

LIQUID CRYSTAL DISPLAYS

In this chapter you will

- Use character LCD modules to display text and numeric data
- Create custom characters to display on character LCD modules
- Use large graphic LCD modules to display text and data
- Create a temperature history graphing thermometer display

For some projects, you'll want to display information to the user somewhere other than a desktop computer monitor. One of the easiest and most versatile ways to display information is with a liquid crystal display (LCD) module and your Arduino. You can display text, custom characters, numeric data using a character LCD module, and graphics with a graphic LCD module.

Character LCD Modules

LCD modules that display characters such as text and numbers are the most inexpensive and simplest to use of all LCDs. They can be purchased in various sizes, which are measured by the number of rows and columns of characters they can display. Some include a backlight and allow you to choose the color of the character and the background color. Any LCD with an HD44780- or KS0066-compatible interface and a 5 V backlight should work with your Arduino. The first LCD we'll use is a 16-character–by–2-row LCD module with a backlight, as shown in Figure 7-1.

Figure 7-1: Example LCD with trimpot and header pins

The *trimpot* (the variable resistor for the LCD) has a value of 10 kΩ and is used to adjust the display contrast. The header pins are soldered into the row of holes along the top of the LCD to make insertion into our breadboard straightforward. The holes along the top of the LCD are numbered 1 through 16. Number 1 is closest to the corner of the module and marked as VSS (connected to GND) in the schematic shown in Figure 7-2. If your LCD has a 4.2 V backlight, place a 1N4004 diode in series between Arduino 5V and LCD LED+ pin. We'll refer to this schematic for all of the LCD examples in this book.

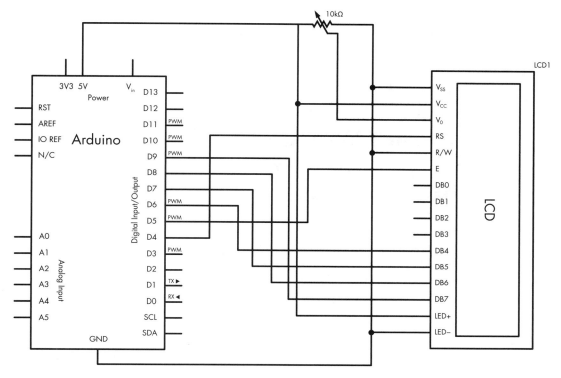

Figure 7-2: Basic LCD schematic

Using a Character LCD in a Sketch

To use the character LCD shown in Figure 7-1, we will first explain the required functions and how they work through some simple demonstrations. Enter and upload the basic sketch shown in Listing 7-1:

```
// Listing 7-1
#include <LiquidCrystal.h>
LiquidCrystal lcd(4, 5, 6, 7, 8, 9);  // pins for RS, E, DB4, DB5, DB6, DB7
void setup()
{
    lcd.begin(16, 2);
    lcd.clear();
}
void loop()
{
  lcd.setCursor(0, 5);
  lcd.print("Hello");
```

```
    lcd.setCursor(1, 6);
    lcd.print("world!");
    delay(10000);
}
```

Listing 7-1: LCD demonstration sketch

Figure 7-3 shows the result of Listing 7-1.

Figure 7-3: LCD demonstration: "Hello world!"

Now to see how the sketch in Listing 7-1 works. First we need to include two initial lines in the sketch. Their purpose is to include the library for LCD modules (which is automatically installed with the Arduino IDE), and then we need to tell the library which pins are connected to the Arduino. To do this, we add the following lines *before* the void setup() method:

```
#include <LiquidCrystal.h>
LiquidCrystal lcd(4, 5, 6, 7, 8, 9);   // pins for RS, E, DB4, DB5, DB6, DB7
```

If you need to use different digital pins on the Arduino, adjust the pin numbers in the second line of this code.

Next, in void setup(), we tell the Arduino the size of the LCD in columns and rows. For example, here's how we'd tell the Arduino that the LCD has two rows of 16 characters each:

```
    lcd.begin(16, 2);
```

Displaying Text

With the LCD setup complete, clear the LCD's display with the following:

```
    lcd.clear();
```

Then, to position the cursor, which is the starting point for the text, use this:

```
    lcd.setCursor(x, y);
```

Here, x is the column (0 to 15) and y is the row (0 or 1). So, for example, to display the word *text*, you would enter the following:

```
    lcd.print("text");
```

Now that you can position and locate text, let's move on to displaying variable data.

Displaying Variables or Numbers

To display the contents of variables on the LCD screen, use this line:

```
lcd.print(variable);
```

If you are displaying a float variable, you can specify the number of decimal places to use. For example, lcd.print(pi, 3) in the following example tells the Arduino to display the value of pi to three decimal places, as shown in Figure 7-4:

```
float pi = 3.141592654;
lcd.print("pi: ");
lcd.print(pi, 3);
```

Figure 7-4: LCD displaying a floating-point number

When you want to display an integer on the LCD screen, you can display it in hexadecimal or binary, as shown in Listing 7-2.

```
// Listing 7-2
  int zz = 170;
  lcd.setCursor(0, 0);
  lcd.print("Binary: ");
  lcd.print(zz, BIN);      // display 170 in binary
  lcd.setCursor(0, 1);
  lcd.print("Hexadecimal: ");
  lcd.print(zz, HX);       // display 170 in hexadecimal
```

Listing 7-2: Functions for displaying binary and hexadecimal numbers

The LCD will then display the text shown in Figure 7-5.

Figure 7-5: Results of the code in Listing 7-2

Project #25: Defining Custom Characters

In addition to using the standard letters, numbers, and symbols available on your keyboard, you can define up to eight of your own characters in each sketch. Notice in the LCD module that each character is made up of eight rows of five dots, or *pixels*. Figure 7-6 shows a close-up.

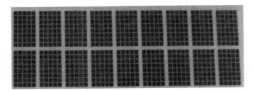

Figure 7-6: Each character is made up of eight rows of five pixels.

To display your own characters, you must first define each one using an *array*. For example, to create a smiley face, you could use the following:

```
byte a[8] = {  B00000,
               B01010,
               B01010,
               B00000,
               B00100,
               B10001,
               B01110,
               B00000  };
```

Each number in the array addresses an individual pixel in the display. A 0 turns off a pixel, and a 1 turns it on. The elements in the array represent the rows of pixels in the display; the top element is the top row, the next element is the second row down, and so on.

NOTE   *When you're planning your custom characters, it can be helpful to plan the character using some graph paper. Each square that is filled in represents a 1, and each empty square represents a 0 in the array.*

In this example, since the first element is B00000, all the pixels in the top row are turned off because we see only 0s. In the next element, B01010, every other pixel is turned on, and the 1s form the top of the eyes. Each row and pixel continues to fill out the characters.

Next, assign the array (which defines your new character) to the first of the eight custom character slots in void setup() with the following function:

```
lcd.createChar(0, a); // assign the array a[8] to custom character slot 0
```

Finally, to display the character, add the following in void loop():

```
lcd.write(byte(0));
```

To display our custom character, we'd use the following code:

```
// Project 25 - Defining Custom Characters
#include <LiquidCrystal.h>
LiquidCrystal lcd(4, 5, 6, 7, 8, 9);   // pins for RS, E, DB4, DB5, DB6, DB7
byte a[8] = {  B00000,
               B01010,
               B01010,
               B00000,
               B00100,
               B10001,
               B01110,
               B00000  };
void setup()
{
   lcd.begin(16, 2);
   lcd.createChar(0, a);
}
void loop()
{
   lcd.write(byte(0));   // write the custom character 0 to the next cursor
                         // position
}
```

Figure 7-7 shows the smiley faces displayed on the LCD screen.

Figure 7-7: The result of Project 25

Character LCD modules are simple to use and somewhat versatile. For example, using what you've learned, you could create a detailed digital thermometer by combining this LCD and the temperature-measurement part of Project 20. However, if you need to display a lot of data or graphical items, you will need to use a *graphic LCD module*.

Graphic LCD Modules

Graphic LCD modules are larger and more expensive than character modules, but they're also more versatile. You can use them not only to display text but also to draw lines, dots, circles, and more to create visual effects. The graphic LCD used in this book is a 128-by-64-pixel module with a KS0108B-compatible interface, as shown in Figure 7-8.

As with the character LCD, the graphic LCD's trimpot has a value of 10 kΩ and is used to adjust the display contrast. The header pins are soldered to the row of holes along the bottom of the LCD to make insertion into a breadboard more convenient, with the holes numbered 1 through 20. The pin closest to the corner of the module is pin 1.

Figure 7-8: A graphic LCD module

Connecting the Graphic LCD

Before you can use the graphic LCD, you'll need to connect 20 wires between the LCD and the Arduino. Make the connections as shown in Table 7-1.

Table 7-1: Graphic LCD-to-Arduino Connections

LCD Pin Number	To Arduino Pin	LCD Pin Function
1	5 V	VDD
2	GND	VSS (GND)
3	Center of 10 kΩ trimpot	VO (Contrast)
4	D8	DB0
5	D9	DB1
6	D10	DB2
7	D11	DB3
8	D4	DB4
9	D5	DB5
10	D6	DB6
11	D7	DB7
12	A0	CS2
13	A1	CS1
14	RST	/RESET
15	A2	R/W
16	A3	D/I
17	A4	E
18	Outer leg of trimpot, other trimpot leg to 5 V	VEE (trimpot voltage)
19	5 V through a 22 Ω resistor	LED backlight anode (+)
20	GND	LED backlight cathode (–)

Using the LCD

After the LCD is wired up, you can download and install the Arduino library for the graphic LCD module. Download the latest version of the library from *http://code.google.com/p/glcd-arduino/downloads/list/* and install it in the same manner described in "Expanding Sketches with Libraries" on page 169.

Now, to use the LCD, insert the following before the void setup():

```
#include <glcd.h>               // include the graphics LCD library
#include "fonts/SystemFont5x7.h" // include the standard character fonts for it
```

Then, after void setup(), add the following lines to prepare the display:

```
GLCD.Init(NON_INVERTED);    // use INVERTED to invert pixels being on or off
GLCD.SelectFont(System5x7); // choose a font for use on the display
GLCD.ClearScreen();         // clear the LCD screen
```

Controlling the Display

The graphic LCD's display will show up to eight rows of 20 characters of text. To position the text cursor, enter the following, replacing the x and y with actual coordinates:

```
GLCD.CursorTo(x, y);
```

To display particular text, replace *text* in the following example with the text you want to display:

```
GLCD.Puts("text");
```

To display an integer, use the following code, replacing *number* with the number you want to display:

```
GLCD.PrintNumber(number);
```

Project #26: Seeing the Text Functions in Action

The text functions are demonstrated in this sketch:

```
// Project 26 - Seeing the Text Functions in Action
#include <glcd.h>               // include the graphics LCD library
#include "fonts/SystemFont5x7.h" // include the standard character fonts for it
```

```
int j = 7;

void setup()
{
  GLCD.Init(NON_INVERTED);
  GLCD.ClearScreen();
  GLCD.SelectFont(System5x7);
}

void loop()
{
  GLCD.ClearScreen();
  GLCD.CursorTo(1, 1);
  GLCD.Puts("Hello, world.");
  GLCD.CursorTo(1, 2);
  GLCD.Puts("I hope you are ");
  GLCD.CursorTo(1, 3);
  GLCD.Puts("enjoying this");
  GLCD.CursorTo(1, 4);
  GLCD.Puts("book. ");
  GLCD.CursorTo(1, 5);
  GLCD.Puts("This is from ");
  GLCD.CursorTo(1, 6);
  GLCD.Puts("chapter ");
  GLCD.PrintNumber(j);
  GLCD.Puts(".");
  do {} while (1);
}
```

The sketch should display the output shown in Figure 7-9.

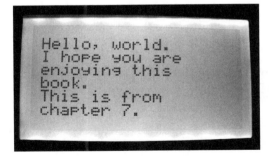

Figure 7-9: Output from Project 26

Creating More Complex Display Effects

Now let's look at a few functions we can use to create various display effects. Keep in mind that the graphic LCD screen has a resolution of 128 columns of 64 pixels, but when we refer to them in various functions, they are counted from 0 to 127 across and 0 to 63 down.

This graphical function will turn on a single pixel at position *x, y* with the color set to BLACK, or it will turn off a single pixel with the color set to WHITE. The color parameter always sets black to on and white to off.

```
GLCD.SetDot(x, y, color); // color is BLACK or WHITE
```

The next function draws a rectangle with the upper-left corner at *x, y*. The width is *w*, and the depth or vertical size is *h*. The resulting rectangle will have a black outline and a white background.

```
GLCD.DrawRect(x, y, w, h, color);
```

This function draws a filled rectangle with the same parameters:

```
GLCD.FillRect(x, y, w, h, color);
```

This function draws a rectangle with the same parameters but with rounded corners of radius *r*.

```
GLCD.DrawRoundRect(x, y, w, h, r, color);
```

This function draws a circle with the center at *x, y* and a radius of *r* pixels:

```
GLCD.DrawCircle(x, y, r, color);
```

This draws a vertical line starting from point *x, y* with a length of 1 pixel.

```
GLCD.DrawVertLine(x, y, l, color);
```

And this draws a horizontal line starting from point *x, y* with a length of 1 pixel.

```
GLCD.DrawHoriLine(x, y, l, color);
```

With the functions discussed so far and some imagination, you can create a variety of display effects or display data graphically. In the next section, we'll build on our quick-read thermometer example using the LCD screen and some of these functions.

Project #27: Creating a Temperature History Monitor

In this project, our goal is to measure the temperature once every 20 minutes and display the last 100 readings in a line graph. Each reading will be represented as a pixel, with the temperature on the vertical axis and time on the horizontal. The most current reading will appear on the left, and the display will continually scroll the readings from left to right. The current temperature will also be displayed as a numeral.

The Algorithm

Although it may sound complex, this project is fairly easy and actually requires only two functions. The first function takes a temperature reading from the TMP36 temperature sensor and stores it in an array of 100 values. Each time a new reading is taken, the previous 99 readings are moved down the array to make way for the new reading, and the oldest reading is erased. The second function draws on the LCD screen. It displays the current temperature, a scale for the graph, and the positions of each pixel for the display of the temperature data over time.

The Hardware

Here's what you'll need to create this project:

- One 128-by-64-pixel KS0108B graphic LCD module with pins for breadboard use
- One 10 kΩ trimpot
- One TMP36 temperature sensor
- Various connecting wires
- One breadboard
- Arduino and USB cable

Connect the graphic LCD as described in Table 7-1, and connect the TMP36 sensor to 5 V, analog 5, and GND.

The Sketch

Enter and upload the following sketch, which also includes relevant comments about the functions used.

```
// Project 27 - Creating a Temperature History Monitor

#include <glcd.h>                 // include the graphics LCD library
#include <fonts/SystemFont5x7.h> // include the standard character fonts for it

int tcurrent;
int tempArray[100];

void setup()
{
  GLCD.Init(NON_INVERTED); // configure GLCD
  GLCD.ClearScreen();      // turn off all GLCD pixels
  GLCD.SelectFont(System5x7);
}
```

```
void getTemp() // function to read temperature from TMP36
{
  float sum = 0;
  float voltage = 0;
  float sensor = 0;
  float celsius;

  // read the temperature sensor and convert the result to degrees C
  sensor   = analogRead(5);
  voltage  = (sensor * 5000) / 1024;
  voltage  = voltage - 500;
  celsius  = voltage / 10;
  tcurrent = int(celsius);

  // insert the new temperature at the start of the array of past temperatures
  for (int a = 99 ; a >= 0 ; --a )
  {
    tempArray[a] = tempArray[a-1];
  }
  tempArray[0] = tcurrent;
}

void drawScreen() // generate GLCD display effects
{
  int q;
  GLCD.ClearScreen();
  GLCD.CursorTo(5, 0);
  GLCD.Puts("Current:");
  GLCD.PrintNumber(tcurrent);
  GLCD.CursorTo(0, 1);
  GLCD.PrintNumber(40);
  GLCD.CursorTo(0, 2);
  GLCD.PrintNumber(32);
  GLCD.CursorTo(0, 3);
  GLCD.PrintNumber(24);
  GLCD.CursorTo(0, 4);
  GLCD.PrintNumber(16);
  GLCD.CursorTo(1, 5);
  GLCD.PrintNumber(8);
  GLCD.CursorTo(1, 6);
  GLCD.PrintNumber(0);
  for (int a = 28 ; a < 127 ; a++)
  {
    q = (55 - tempArray[a-28]);
    GLCD.SetDot(a, q, BLACK);
  }
}

void loop()
{
  getTemp();
  drawScreen();
  for (int a = 0 ; a < 20 ; a++) // wait 20 minutes until the next reading
```

```
    {
        delay(60000);          // wait 1 minute
    }
}
```

The Result

The resulting display should look something like Figure 7-10.

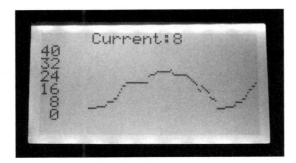

Figure 7-10: Results of Project 27

Modifying the Sketch

Some people can better interpret data in a visual way instead of just reading numbers. This type of project could also be used to display other kinds of data, such as the voltage from various sensors as measured by analog input pins. Or you could add another temperature sensor and show both values at once. Almost anything that returns a value can be displayed using the graphic LCD module.

Looking Ahead

Now that you have experience with LCDs, you can see that the Arduino is in fact a small computer: It can accept and process incoming data and display it to the outside world. But this is only the beginning. In the next chapter you'll work on making your own Arduino protoshields, record data to microSD memory cards, and learn about libraries and Arduino timing functions.

8

EXPANDING YOUR ARDUINO

In this chapter you will

- Learn about the broad variety of Arduino shields
- Make your own Arduino shield using a ProtoShield
- Understand how Arduino libraries can expand the available functions
- Use a microSD card shield to record data that can be analyzed in a spreadsheet
- Build a temperature-logging device
- Learn how to make a stopwatch using `micros()` and `millis()`
- Understand Arduino interrupts and their uses

We'll continue to discover ways to expand our Arduino (using various shields), and you'll follow an example to learn how to make your own shield. Over time, as you experiment with electronics and Arduino, you can make your circuits more permanent by building them onto a *ProtoShield*, a blank printed circuit board that you can use to mount custom circuitry.

One of the more useful shields is the *microSD* card shield. We'll use it in this chapter to create a temperature-logging device to record temperatures over time; the shield will be used to record data from the Arduino to be transferred elsewhere for analysis.

You'll learn about the functions `micros()` and `millis()`, which are very useful for keeping time, as you'll see in the stopwatch project. Finally, we'll examine interrupts.

Shields

You can add functionality to your Arduino board by attaching *shields*. A shield is a circuit board that connects via pins to the sockets on the sides of an Arduino. Hundreds of shields are available on the market. One popular project, for example, combines a GPS shield with a microSD memory card shield to create a device that logs and stores position over time, such as a car's path of travel or the location of a new hiking trail. Other projects include Ethernet network adapters that let the Arduino access the Internet (Figure 8-1).

Figure 8-1: Ethernet shield on Arduino Uno

GPS satellite receivers let you track the location of the Arduino (Figure 8-2). MicroSD memory card interfaces let the Arduino store data on a memory card (Figure 8-3).

Arduino shields are designed to be stacked, and they usually work in combination with other shields. For example, Figure 8-4 shows a stack that includes an Arduino Uno, a microSD memory card shield to which data can be recorded, an Ethernet shield for connecting to the Internet, and an LCD shield to display information.

Figure 8-2: GPS receiver shield (with separate GPS module) on Arduino Uno

Figure 8-3: MicroSD card shield kit

Figure 8-4: Three stacked shields with an Arduino Uno

ProtoShields

You can buy a variety of shields online (at *http://www.shieldlist.org/*, for example) or make your own using a ProtoShield. ProtoShields come pre-assembled or in kit form, similar to the one shown in Figure 8-5.

A ProtoShield also makes a good base for a solderless breadboard, because it keeps a small circuit within the physical boundary of your Arduino creation (as with the quick-read thermometer shown in Figure 8-6). Smaller solderless breadboards fit within the rows of sockets can be attached to the circuit board with Blu-Tack reusable putty for temporary mounting or double-sided tape for more permanent use. ProtoShields can also act as a more permanent foundation for circuits that have been tested on a breadboard.

Figure 8-5: ProtoShield kit

Figure 8-6: The quick-read thermometer from Project 8

Building custom circuits on a ProtoShield requires strategic and special planning. This includes designing the circuit, making a schematic, and then planning the layout of the components as they will sit in the ProtoShield. Finally, the completed circuit for your custom shield will be soldered into place, but you should always test it first using a solderless breadboard to ensure that it works. Some ProtoShields come with a PDF schematic file that you can download and print, intended specifically for drawing your project schematic.

Project #28: Creating a Custom Shield with Eight LEDs

In this project, you'll create a custom shield containing eight LEDs and current-limiting resistors. This custom shield will make it easy to experiment with LEDs on digital outputs.

The Hardware

The following hardware is required for this project:

- One blank Arduino ProtoShield
- Eight LEDs of any color
- Eight 560 Ω resistors (R1 to R8)
- Two 6-pin Arduino stackable headers
- Two 8-pin Arduino stackable headers

The Schematic

The circuit schematic is shown in Figure 8-7.

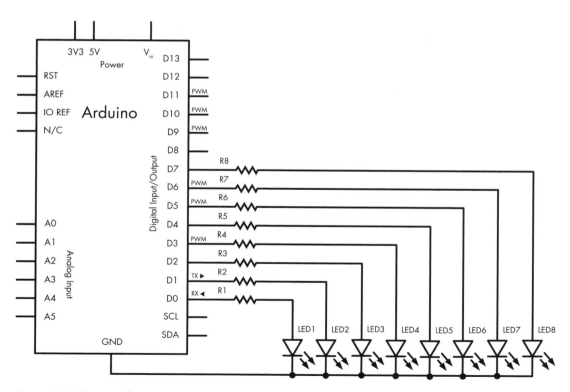

Figure 8-7: Schematic for Project 28

The Layout of the ProtoShield Board

The next step is to learn the layout of the holes on the ProtoShield. The rows and columns of holes on the ProtoShield should match those of the Arduino's solderless board. On the blank ProtoShield shown in Figure 8-8, the designers have surrounded holes that are electrically connected with black lines.

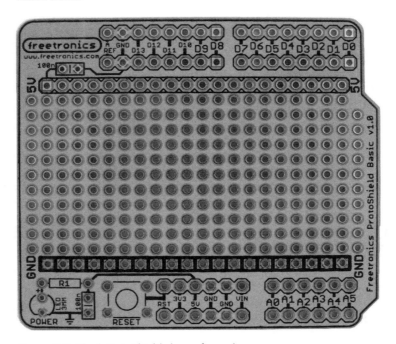

Figure 8-8: Blank ProtoShield shown from above

The long horizontal line below the digital pins is connected to the 5V socket, and the horizontal line above the analog and other pins is connected to GND. The holes between the two lines are insulated from each other. Finally, there are two rows of pins: one at the very top of the ProtoShield and one at the very bottom. The top row consists of two groups of eight pins, and the bottom has two groups of six pins. This is where we solder the stackable headers that allow the ProtoShield to slot into the Arduino board.

The Design

Lay out your circuit using graph paper, as shown in Figure 8-9.

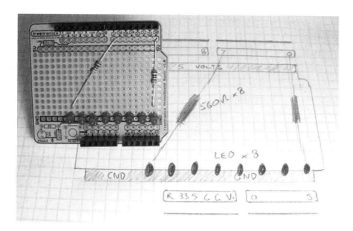

Figure 8-9: Planning our custom shield

After you've drawn a plan for your circuit, test-fit the actual components into the ProtoShield to make sure that they'll fit and that they aren't too crowded. If the ProtoShield has space for a reset button, always include one, because the shield will block access to your Arduino's RESET button.

Soldering the Components

Once you're satisfied with the layout of the circuit on your ProtoShield and you have tested the circuit to make sure that it works, you can solder the components. Using a soldering iron is not that difficult, and you don't need to buy an expensive soldering station for this type of work. A simple iron rated at 25–40 watts, like the one shown in Figure 8-10, should do it.

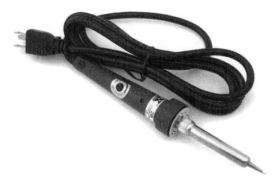

Figure 8-10: Soldering iron

NOTE　*If soldering is new to you, download and read the instructional comic book from* http://mightyohm.com/soldercomic/.

When soldering the components, you may need to bridge them together with a small amount of solder, as shown in Figure 8-11. As you can see, the end of one resistor is connected to the anode of an LED.

Check each solder connection as you go, because mistakes are easier to locate and repair *before* the project is finished. When the time comes to solder the four header sockets, keep them aligned using an existing shield to hold the new sockets, as shown in Figure 8-12.

Figure 8-11: Solder bridge

Figure 8-12: Soldering header sockets

Figure 8-13 shows the finished product: a custom Arduino shield with eight LEDs.

Figure 8-13: Completed custom shield!

Modifying the Custom Shield

We could use this simple shield in a more complicated project—for example, to monitor the status of digital pins 0 to 7. If we added another six resistors and LEDs, then we could monitor the entire digital output range. There are lots of different ways to use this shield. Just use your imagination!

Expanding Sketches with Libraries

Just as an Arduino shield can expand our hardware, a *library* can add useful functions for particular tasks that we can add to our sketches or add extra functions that allow the use of various hardware specific to a manufacturer. Anyone can create a library, just as suppliers of various Arduino shields often write their own libraries to match their hardware.

The Arduino IDE already includes a set of preinstalled libraries. To include them in your sketches, choose **Sketch ▸ Import Library**, and you should see the collection of preinstalled libraries with names such as Ethernet, LiquidCrystal, Servo, and so on. Many of these names will be self-explanatory. (If a library is required in your project work, it will be explained in detail.)

Importing a Shield's Libraries

If you buy a shield, you'll generally need to download and install its libraries from the shield vendor's site or from a link provided.

To demonstrate how this is done, let's download the library required by the microSD card shield shown in Figure 8-3.

1. Visit *https://github.com/greiman/SdFat/* and click the **Download ZIP** button. Figure 8-14 shows this web page.

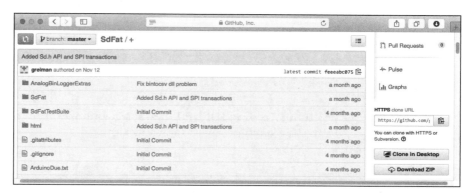

Figure 8-14: Library download page

2. After a moment, the file *SdFat-master.zip* will appear in your *Downloads* folder. Double-click to unzip this folder, which reveals the contents as shown in Figure 8-15.

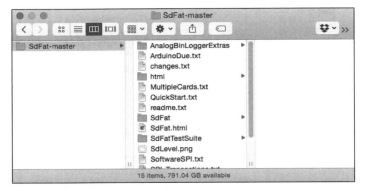

Figure 8-15: Library folder

3. Once you've downloaded the library, use the directions that follow to install it on your operating system.

Installing the Library on Mac OS X

If you're downloading to the Mac OS X system, follow these directions:

1. Open the *Downloads* folder and find the downloaded file folder, as shown in Figure 8-16.

Figure 8-16: Downloaded library folder

2. Open the Arduino IDE and then select **Import Library ▸ Add Library** from the Sketch menu, as shown in Figure 8-17.

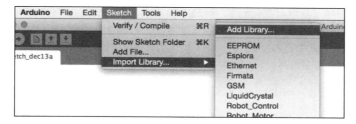

Figure 8-17: Importing the library

3. When the dialog shown in Figure 8-18 appears, select the *SdFat-master* folder located in the *Downloads* folder and click **Choose**.

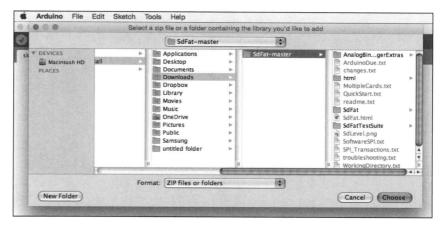

Figure 8-18: Choosing the library folder

4. Make sure that the *SdFat* library has been installed and is available by restarting the IDE software and selecting **Sketch ▸ Import Library**. *SdFat* should show up in the list. (If it does not, try the installation procedure again.)

With your library installed, skip forward to "MicroSD Memory Cards" on page 173.

Installing the Library on Windows XP and Later

If you're downloading to a Windows XP or later system, follow these directions:

1. After the download has completed, open the downloaded file folder to reveal its contents, and then locate the *SdFat* folder, as shown in Figure 8-19.

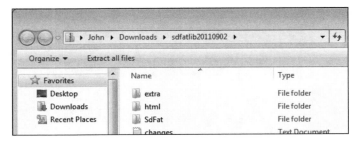

Figure 8-19: Downloaded library folder

2. Copy the *SdFat* folder from the window shown in Figure 8-19 to your *Arduino/libraries* folder, as shown in Figure 8-20.

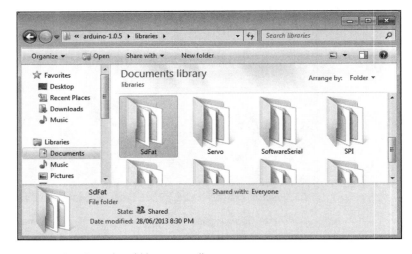

Figure 8-20: Completed library installation

3. Make sure that the *SdFat* library has been installed and is available by restarting the IDE and selecting **Sketch ▸ Import Library**. *SdFat* should appear in the list. (If it does not, try the installation procedure again from the beginning.)

    With your library installed, skip forward to "MicroSD Memory Cards" on page 173.

Installing the Library on Ubuntu Linux 11.04 and Later

If you're downloading to a system running Ubuntu Linux 11.04 or later, follow these directions:

1. Locate the downloaded file and double-click it. The Archive Manager window should appear with the *SdFat* folder, as shown in Figure 8-21.

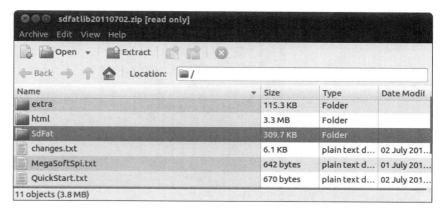

Figure 8-21: Downloaded library folder

2. Right-click the *SdFat* folder in the window shown in Figure 8-21, and then click **Extract** to extract it to your */libraries* folder (Figure 8-22).

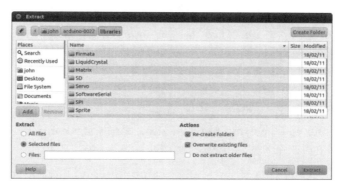

Figure 8-22: Destination for library folder

3. Make sure that the *SdFat* library has been installed and is available by restarting the IDE and selecting **Sketch ▶ Import Library**. *SdFat* should appear in the list. (If it does not, try the installation procedure again from the beginning.)

With your library installed, move on to the next section.

MicroSD Memory Cards

By using microSD cards with your Arduino, you can capture data from many sources, such as the TMP36 temperature sensor we used in Chapter 4. You can also use the microSD card to store web server data or any files for your project to use. To record and store the data you collect, you can use a microSD memory card shield like the one shown in Figure 8-23. This shield uses microSD (not microSDHC) cards with a capacity of up to 2GB.

Figure 8-23: A microSD card with a 2GB capacity

Several memory card shields are available from popular retailers such as SparkFun, Adafruit Industries, and Snootlab. The shield used in this book (shown in Figure 8-3) is from SparkFun (part numbers DEV-09802 and PRT-10007).

NOTE   *Before you can use a memory card with a shield, you need to format it. To do so, plug it into a computer and follow your operating system's instructions for formatting memory cards. Make sure you use the FAT16 format type. Also, depending on which shield you buy, the sockets may need to be soldered in, following the procedure discussed in "Soldering the Components" on page 167.*

Testing Your MicroSD Card

After you have finished formatting and assembling the microSD card shield, make sure that it's working correctly. To do so, follow these steps:

1. Connect the shield to your Arduino, and then insert the memory card and plug in the USB cable.
2. Run the IDE by selecting **File ▶ Examples ▶ SdFat ▶ SdInfo**. Then upload the *SdInfo* sketch that appeared in the IDE.
3. Open the Serial Monitor window, set it to 9,600 baud, press any key on the keyboard, and then press ENTER. After a moment, you should see some data about the microSD card, as shown in Figure 8-24.

If the test results don't appear in the Serial Monitor, then try the following until the problem is fixed:

*   Remove the USB cable from your Arduino, and remove and reinsert the microSD card.
*   Make sure that the header sockets are soldered neatly and that the pins are not shorted out.
*   Check that the Serial Monitor baud rate is 9,600 and that a regular Arduino Uno–compatible board is being used. The Mega and some other board models have the SPI pins in different locations.
*   Reformat your microSD card.

Figure 8-24: Successful microSD card test results

Project #29: Writing Data to the Memory Card

To write data to the memory card, connect your shield, insert a microSD card, and then enter and upload the following sketch:

```
// Project 29 - Writing Data to the Memory Card

int b = 0;

#include <SD.h>
void setup()
{
  Serial.begin(9600);
  Serial.print("Initializing SD card...");
  pinMode(10, OUTPUT);

  // check that the microSD card exists and is usable
  if (!SD.begin(8))
  {
    Serial.println("Card failed, or not present");
    // stop sketch
    return;
  }
  Serial.println("microSD card is ready");
}

void loop()
{
```

```
❶    // create the file for writing
     File dataFile = SD.open("DATA.TXT", FILE_WRITE);
     // if the file is ready, write to it:
     if (dataFile)
❷    {
       for ( int a = 0 ; a < 11 ; a++ )
       {
         dataFile.print(a);
         dataFile.print(" multiplied by two is ");
         b = a * 2;
❸        dataFile.println(b, DEC);
❹      }
❺      dataFile.close(); // close the file once the system has finished with it
                         // (mandatory)
     }
     // if the file isn't ready, show an error:
     else
     {
       Serial.println("error opening DATA.TXT");
❻    }
     Serial.println("finished");
     do {} while (1);
}
```

The sketch creates a text file called *DATA.txt* on the microSD card, as shown in Figure 8-25.

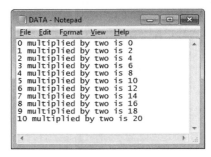

Figure 8-25: Output from Project 29

Let's review the void loop() section of the sketch to see how it created the text file. The code in void loop() between ❶ and ❷ and between ❹ and ❻ creates and opens the file for writing. To write text to the file, we use this:

```
dataFile.print(); or dataFile.println();
```

This code works in the same manner as, for example, Serial.println(), so you can write it in the same manner as you would to the Serial Monitor. At ❶, we can set the name of the created text file, which must be eight characters, followed by a dot, and then three characters, such as *DATA.txt*.

At ❸, we use DEC as the second parameter. This states that the variable is a decimal number and should be written to the text file as such. If we were writing a float variable instead, then we would use a digit for the number of decimal places to write (to a maximum of six).

When finished writing data to the file, at ❺, we use dataFile.close() to close the file for writing. If this step is not followed, the computer will not be able to read the created text file.

Project #30: Creating a Temperature-Logging Device

Now that you know how to record data, let's measure the temperature every minute for 8 hours using our microSD card shield and the TMP36 temperature sensor that we introduced in Chapter 4. We'll combine the functions for writing to the microSD card from Project 29 with temperature measurement from Project 27.

The Hardware

The following hardware is required:

- One TMP36 temperature sensor
- One breadboard
- Various connecting wires
- MicroSD card shield and card
- Arduino and USB cable

Insert the microSD card into the shield, and then insert the shield into the Arduino. Connect the left (5V) pin of the TMP36 to Arduino 5V, the middle pin to analog, and the right pin to GND.

The Sketch

Enter and upload the following sketch:

```
// Project 30 - Creating a Temperature-Logging Device

#include <SD.h>
float sensor, voltage, celsius;

void setup()
{
  Serial.begin(9600);
  Serial.print("Initializing SD card...");
  pinMode(10, OUTPUT);

  // check that the microSD card exists and can be used
  if (!SD.begin(8))
  {
    Serial.println("Card failed, or not present");
```

```
    // stop sketch
    return;
  }
  Serial.println("microSD card is ready");
}
void loop()
{
  // create the file for writing
  File dataFile = SD.open("DATA.TXT", FILE_WRITE);
  // if the file is ready, write to it:
  if (dataFile)
  {
    for ( int a = 0 ; a < 481 ; a++ ) // 480 minutes in 8 hours
    {
      sensor = analogRead(0);
      voltage = (sensor * 5000) / 1024; // convert raw sensor value to
                                        // millivolts
      voltage = voltage - 500;
      celsius = voltage / 10;
      dataFile.print(" Log: ");
      dataFile.print(a, DEC);
      dataFile.print(" Temperature: ");
      dataFile.print(celsius, 2);
      dataFile.println(" degrees C");
      delay(599900); // wait just under one minute
    }
    dataFile.close(); // mandatory
    Serial.println("Finished!");
    do {} while (1);
  }
}
```

The sketch will take a little more than 8 hours to complete, but you can alter this time period by lowering the value in delay(599900).

After the sketch has finished, remove the microSD card and open the log file in a text editor, as shown in Figure 8-26.

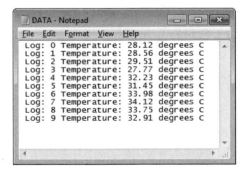

Figure 8-26: Results from Project 30

For more serious analysis of the captured data, delimit the lines of text written to the log file with spaces or colons so that the file can be easily imported into a spreadsheet. For example, you could import the file into OpenOffice Calc by choosing **Insert ▸ Sheet From File** to produce a spreadsheet like the one shown in Figure 8-27, which you could then analyze the data of, as shown in Figure 8-28.

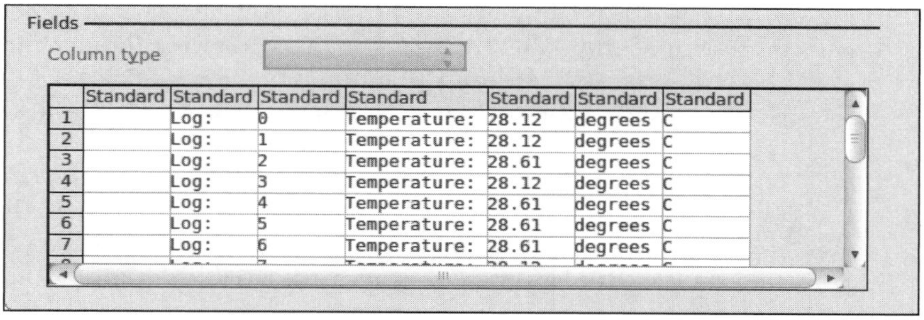

	Standard	Standard	Standard	Standard	Standard	Standard	Standard
1		Log:	0	Temperature:	28.12	degrees	C
2		Log:	1	Temperature:	28.12	degrees	C
3		Log:	2	Temperature:	28.61	degrees	C
4		Log:	3	Temperature:	28.12	degrees	C
5		Log:	4	Temperature:	28.61	degrees	C
6		Log:	5	Temperature:	28.61	degrees	C
7		Log:	6	Temperature:	28.61	degrees	C

Figure 8-27: Importing data into a spreadsheet

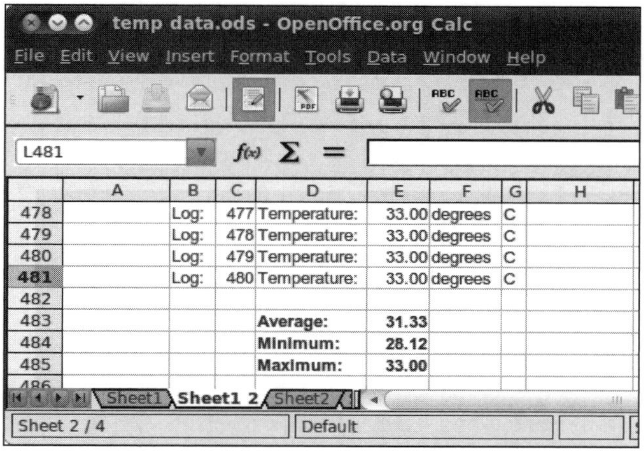

	A	B	C	D	E	F	G	H
478		Log:	477	Temperature:	33.00	degrees	C	
479		Log:	478	Temperature:	33.00	degrees	C	
480		Log:	479	Temperature:	33.00	degrees	C	
481		Log:	480	Temperature:	33.00	degrees	C	
482								
483				**Average:**	**31.33**			
484				**Minimum:**	**28.12**			
485				**Maximum:**	**33.00**			

Figure 8-28: Temperature analysis

The temperature examples can be hacked to suit your own data analysis projects. You can use these same concepts to record any form of data that can be generated by an Arduino system.

Timing Applications with millis() and micros()

Each time the Arduino starts running a sketch, it also records the passage of time using milliseconds and microseconds. A millisecond is one thousandth of a second (0.001), and a microsecond is one millionth of a second (0.000001). You can use these values to measure the passage of time when running sketches.

The following functions will access the time values stored in an unsigned long variable:

```
unsigned long a,b;
a = micros();
b = millis();
```

Due to the limitations of the unsigned long variable type, the value will reset to 0 after reaching 4,294,967,295, allowing for around 50 days of counting for millis() and 70 minutes in micros(). Furthermore, due to the limitations of the Arduino's microprocessor, micros() values are always in multiples of four.

Let's use these values to see how long it takes for the Arduino to turn a digital pin from LOW to HIGH and vice versa. To do this, we'll read micros() before and after a digitalWrite() function, find the difference, and display it on the Serial Monitor. The only required hardware is your Arduino and cable.

Enter and upload the sketch shown in Listing 8-1.

```
// Listing 8-1

unsigned long start, finished, elapsed;

void setup()
{
  Serial.begin(9600);
  pinMode(3, OUTPUT);
  digitalWrite(3, LOW);
}

void loop()
{
❶  start = micros();
  digitalWrite(3, HIGH);
❷  finished = micros();
❸  elapsed = finished - start;
  Serial.print("LOW to HIGH: ");
  Serial.print(elapsed);
  Serial.println(" microseconds");
  delay(1000);

❹  start = micros();
  digitalWrite(3, LOW);
  finished = micros();
  elapsed = finished - start;
  Serial.print("HIGH to LOW: ");
  Serial.print(elapsed);
  Serial.println(" microseconds");
  delay(1000);
}
```

Listing 8-1: Timing digital pin state change with micros()

The sketch takes readings of `micros()` before and after the `digitalWrite(HIGH)` functions at ❶ and ❷, and then it calculates the difference and displays it on the Serial Monitor at ❸. This is repeated for the opposite function at ❹.

Now open the Serial Monitor to view the results, shown in Figure 8-29.

Because the resolution is 4 microseconds, if the value is 8 microseconds, then we know that the duration is greater than 4 and less than or equal to 8.

Figure 8-29: Output from Listing 8-1

Project #31: Creating a Stopwatch

Now that we can measure the elapsed time between two events, we can create a simple stopwatch using an Arduino. Our stopwatch will use two buttons: one to start or reset the count and another to stop counting and show the elapsed time. The sketch will continually check the status of the two buttons. When the start button is pressed, a `millis()` value will be stored, and when the second button is pressed, a new `millis()` value will be stored. The custom function `displayResult()` will convert the elapsed time from milliseconds into hours, minutes, and seconds. Finally, the time will be displayed on the Serial Monitor.

The Hardware

The following hardware is required for this project:

- One breadboard
- Two push buttons (S1 and S2)
- Two 10 kΩ resistors (R1 and R2)
- Various connecting wires
- Arduino and USB cable

The Schematic

The circuit schematic is shown in Figure 8-30.

NOTE   *You will use this circuit for the next project, so don't pull it apart when you're finished!*

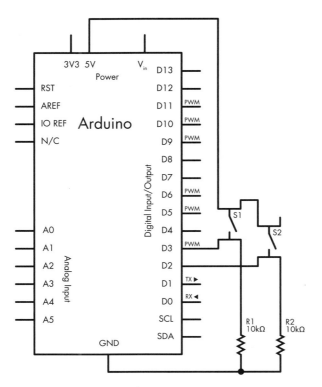

Figure 8-30: Schematic for Project 31

The Sketch

Enter and upload this sketch:

```
// Project 31 - Creating a Stopwatch

unsigned long start, finished, elapsed;

void setup()
{
  Serial.begin(9600);
  pinMode(2, INPUT); // the start button
  pinMode(3, INPUT); // the stop button
  Serial.println("Press 1 for Start/reset, 2 for elapsed time");
}

void displayResult()
{
  float h, m, s, ms;
  unsigned long over;

  elapsed = finished - start;

  h    = int(elapsed / 3600000);
  over = elapsed % 3600000;
```

❶

❷

```
m    = int(over / 60000);
over = over % 60000;
s    = int(over / 1000);
ms   = over % 1000;

Serial.print("Raw elapsed time: ");
Serial.println(elapsed);
Serial.print("Elapsed time: ");
Serial.print(h, 0);
Serial.print("h ");
Serial.print(m, 0);
Serial.print("m ");
Serial.print(s, 0);
Serial.print("s ");
Serial.print(ms, 0);
Serial.println("ms");
Serial.println();
}

void loop()
{
❸   if (digitalRead(2) == HIGH)
    {
      start = millis();
      delay(200); // for debounce
      Serial.println("Started...");
    }
❹   if (digitalRead(3) == HIGH)
    {
      finished = millis();
      delay(200); // for debounce
      displayResult();
    }
}
```

The basis for our stopwatch is simple. At ❶, we set up the digital input pins for the start and stop buttons. At ❸, if the start button is pressed, then the Arduino notes the value for millis() that we use to calculate the elapsed time once the stop button is pressed at ❹. After the stop button is pressed, the elapsed time is calculated in the function displayResult() at ❷ and shown in the Serial Monitor window.

The results shown in Figure 8-31 should appear in the Serial Monitor.

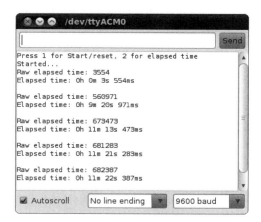

Figure 8-31: Output from Project 31

Interrupts

An *interrupt* in the Arduino world is basically a signal that allows a function to be called at any time within a sketch—for example, when a digital input pin's state changes or a timer event is triggered. Interrupts are perfect for calling a function to interrupt the normal operation of a sketch, such as when a button is pressed.

When an interrupt is triggered, the normal operation and running of your program in the void loop() is halted temporarily, the interrupt function is called and executed, and then when the interrupt function exits, whatever was happening in the main loop starts exactly from where it left off.

Keep in mind that interrupt functions should be short and usually simple. They should exit quickly, and if the interrupt function does something that the main loop may already be doing, then the interrupt function is temporarily going to override whatever the main loop was doing. For example, if the main loop is regularly sending *Hello* out the serial port and the interrupt function sends --- when it is triggered, then you will see any of these come out the serial port: *H----ello, He----llo, Hel----lo, Hell----o,* or *Hello----.*

The Arduino Uno offers two interrupts that are available using digital pins 2 and 3. When properly configured, the Arduino will monitor the voltage applied to the pins. When the voltage changes in a certain, defined way (when a button is pressed, for example), an interrupt is triggered, causing a corresponding function to run—maybe something like "Stop Pressing Me!"

Interrupt Modes

One of four changes (or *modes*) can trigger an interrupt:

- LOW: No current is applied to the interrupt pin.
- CHANGE: The current changes, either between on and off or between off and on.
- RISING: The current changes from off to on at 5 V.
- FALLING: The current changes from on at 5 V to off.

For example, to detect when a button attached to an interrupt pin has been pressed, we could use the RISING mode. Or, for example, if you had an electric trip wire running around your garden (connected between 5 V and the interrupt pin), then you could use the FALLING mode to detect when the wire has been tripped and broken.

NOTE   *The delay() and Serial.available() functions will not work within a function that has been called by an interrupt.*

Configuring Interrupts

To configure interrupts, use the following in void setup():

```
attachInterrupt(0, function, mode);
attachInterrupt(1, function, mode);
```

Here, 0 is for digital pin 2, 1 is for digital pin 3, function is the name of the function to call when the interrupt is triggered, and mode is one of the four modes that triggers the interrupt.

Activating or Deactivating Interrupts

Sometimes within a sketch you won't want to use the interrupts. Deactivate them using the following:

```
noInterrupts(); // deactivate interrupts
```

And then reactivate them with this:

```
interrupts(); // reactivate interrupts
```

Interrupts work quickly and they are very sensitive, which makes them useful for time-critical applications or for "emergency stop" buttons on projects.

Project #32: Using Interrupts

We'll use the circuit from Project 31 to demonstrate the use of interrupts. Our example will blink the built-in LED every 500 milliseconds, during which time both interrupt pins will be monitored. When the button on interrupt 0 is pressed, the value for micros() will be displayed on the Serial Monitor, and when the button on interrupt 1 is pressed, the value for millis() will be displayed.

The Sketch

Enter and upload the following sketch:

```
// Project 32 - Using Interrupts

#define LED 13
void setup()
{
  Serial.begin(9600);
  pinMode(13, OUTPUT);
  attachInterrupt(0, displayMicros, RISING);
  attachInterrupt(1, displayMillis, RISING);
}
```

```
❶ void displayMicros()
  {
    Serial.write("micros() = ");
    Serial.println(micros());
  }

❷ void displayMillis()
  {
    Serial.write("millis() = ");
    Serial.println(millis());
  }

❸ void loop()
  {
    digitalWrite(LED, HIGH);
    delay(500);
    digitalWrite(LED, LOW);
    delay(500);
  }
```

This sketch will blink the onboard LED as shown in void loop() at ❸. When interrupt 0 is triggered, the function displayMicros() at ❶ will be called; or when interrupt 1 is triggered, the function displayMillis() at ❷ will be called. After either function has finished, the sketch resumes running the code in void loop.

Open the Serial Monitor window and press the two buttons to view the values for millis() and micros() as shown in Figure 8-32.

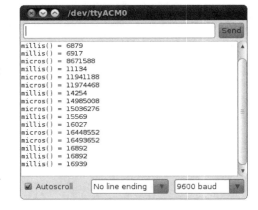

Figure 8-32: Output from Project 32

Looking Ahead

This chapter has given you more tools and options that you can adapt to create and improve your own projects. In future chapters, we will work with more Arduino shields and use the microSD card shield in other data-logging applications.

9

NUMERIC KEYPADS

In this chapter you will

- Learn how to connect numeric keypads to your Arduino
- Read values from the keypad in a sketch
- Expand on decision systems with the switch-case function
- Create a PIN-controlled lock or switch

Using a Numeric Keypad

As your projects become more involved, you might want to accept numeric input from users when your Arduino isn't connected to a PC—for example, if you'd like to have the ability to turn something on or off by entering a secret number. One option would be to wire up 10 or more push buttons to various digital input pins (for the numbers 0 through 9); but it's much easier to use a numeric keypad like the one shown in Figure 9-1.

Figure 9-1: Numeric keypad

One of the benefits of using a keypad is that it uses only 7 pins for 12 active buttons, and with the use of a clever Arduino library, you won't need to add pull-down resistors for debouncing as we did in Chapter 4. You can download and then install the *Keypad* library from *http://arduino.cc/ playground/uploads/Code/Keypad.zip*.

Wiring a Keypad

Physically wiring the keypad to the Arduino is easy. On the back of the keypad are seven pins, as shown in Figure 9-2.

Figure 9-2: Keypad pins

The pins are numbered 1 to 7, from left to right. For all of the keypad projects in this book, you'll make the connections shown in Table 9-1.

Table 9-1: Keypad-to-Arduino Connections

Keypad Pin Number	Arduino Pin
1	Digital 5
2	Digital 4
3	Digital 3
4	Digital 2
5	Digital 8
6	Digital 7
7	Digital 6

Programming for the Keypad

When you write a sketch for the keypad, you must include certain lines of code to enable the keypad, as identified in the following example sketch. The required code starts at ❶ and ends at ❺.

Before moving forward, test the keypad by entering and uploading Listing 9-1:

```
// Listing 9-1

❶ // Beginning of necessary code

  #include "Keypad.h"
  const byte ROWS = 4;          // set display to four rows
  const byte COLS = 3;          // set display to three columns
  char keys[ROWS][COLS] =
    {{'1','2','3'},
     {'4','5','6'},
     {'7','8','9'},
❷   {'*','0','#'}};
❸ byte rowPins[ROWS] = {5, 4, 3, 2};
❹ byte colPins[COLS] = {8, 7, 6};
  Keypad keypad = Keypad( makeKeymap(keys), rowPins, colPins, ROWS, COLS );

❺ // End of necessary code

  void setup()
  {
    Serial.begin(9600);
  }

  void loop()
  {
    char key = keypad.getKey();
    if (key != NO_KEY)
    {
      Serial.print(key);
    }
  }
```

Listing 9-1: Numeric keypad demonstration sketch

At ❷, we introduce the char variable type that contains one character, such as a letter, number, or symbol, that can be generated with a computer keyboard. In this case, it contains the keypad's numbers and symbols. The lines of code at ❸ and ❹ define which pins are used on the Arduino. Using these lines and Table 9-1, you can change the digital pins used for input if you want.

Testing the Sketch

After uploading the sketch, open the Serial Monitor and press some keys on the keypad. The characters for the keys you pressed will be displayed in the Serial Monitor, as shown in Figure 9-3.

Figure 9-3: Results of pressing keys on the keypad

Making Decisions with switch-case

When you need to compare two or more variables, you'll often find it easier to use a switch-case statement instead of an if-then statement, because switch-case statements can make an indefinite number of comparisons and run code when the comparison is found to be true. For example, if we had the integer variable xx with a possible value of 1, 2, or 3 and we wanted to run certain code based on whether a value was 1, 2, or 3, then we could use code like the following to replace our if-then statement:

```
switch(xx)
{
    case 1:
     // do something as the value of xx is 1
    break;              // finish and move on with sketch
    case 2:
    // do something as the value of xx is 2
    break;
    case 3:
    // do something as the value of xx is 3
    break;
    default:
    // do something if xx is not 1, 2 or 3
    // default is optional
}
```

The optional default: section at the end of this code segment lets you choose to run some code when true comparisons no longer exist in the switch-case statement.

Project #33: Creating a Keypad-Controlled Lock

In this project, we'll create the beginning part of a keypad-controlled lock. We'll use the basic setup described in the sketch in Listing 9-1, but we'll also include a secret code that will need to be entered on the keypad. The Serial Monitor will tell the user who types a code into the keypad whether the code is correct or not.

The sketch will call different functions, depending on whether the six-digit secret code is correct. The secret code is stored in the sketch but is not displayed to the user. To activate and deactivate the lock, the user must press * and then the secret number, followed by #.

The Sketch

Enter and upload this sketch:

```
// Project 33 - Creating a Keypad-Controlled Lock

// Beginning of necessary code

#include "Keypad.h"
const byte ROWS = 4;                // set display to four rows
const byte COLS = 3;                // set display to three columns
char keys[ROWS][COLS] =
  {{'1','2','3'},
   {'4','5','6'},
   {'7','8','9'},
   {'*','0','#'}};
byte rowPins[ROWS] = {5, 4, 3, 2};
byte colPins[COLS] = {8, 7, 6};
Keypad keypad = Keypad( makeKeymap(keys), rowPins, colPins, ROWS, COLS );

// End of necessary code
```
❶
```
char PIN[6]={'1','2','3','4','5','6'}; // our secret number
char attempt[6]={0,0,0,0,0,0};
int z=0;

void setup()
{
  Serial.begin(9600);
}

void correctPIN()  // do this if the correct PIN is entered
{
  Serial.println("Correct PIN entered...");
}

void incorrectPIN() // do this if an incorrect PIN is entered
{
  Serial.println("Incorrect PIN entered!");
}

void checkPIN()
{
  int correct=0;
  int i;
```
❷
```
  for ( i = 0; i < 6 ; i++ )
  {
```

```
          if (attempt[i]==PIN[i])
          {
            correct++;
          }
        }
        if (correct==6)
        {
❸         correctPIN();
        } else
        {
❹         incorrectPIN();
        }
        for (int zz=0; zz<6; zz++) // remove previously entered code attempt from
        {
          attempt[zz]=0;
        }
      }

      void readKeypad()
      {
        char key = keypad.getKey();
        if (key != NO_KEY)
        {
❺         switch(key)
          {
          case '*':
            z=0;
            break;
          case '#':
            delay(100);  // removes the possibility of switch bounce
            checkPIN();
            break;
          default:
            attempt[z]=key;
            z++;
          }
        }
      }

      void loop()
      {
❻       readKeypad();
      }
```

How It Works

After the usual setup routines (as described in Listing 9-1), the sketch continually "listens" to the keypad by running the function readKeypad() at ❻. After a key is pressed, the value of the key is examined using a switch-case statement at ❺. The values of the keys pressed on the keypad are stored in the array attempt[], and when the user presses #, the function checkPin() is called.

At ❷, the values of keys pressed are compared against the PIN stored in the array PIN[] at ❶, which holds the secret number. If the correct sequence is entered, the function correctPin() at ❸ is called, where you can add your own code to execute; but if the incorrect sequence is entered, the function incorrectPin() is called at ❹. Finally, once the user's entry has been checked, it is removed from memory and the code is ready for the next test.

Testing the Sketch

After you've uploaded the sketch to the Arduino, open the Serial Monitor window, press the asterisk key (*) on the numeric keypad, type the secret number, and then press the pound sign key (#) when you've finished. Try entering both correct and incorrect numbers. Your results should be similar to the output shown in Figure 9-4.

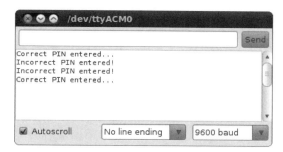

Figure 9-4: Results from entering correct and incorrect PINs

This example served as a perfect foundation for your own PIN-activated devices, such as locks, alarms, or anything else you can imagine. Just be sure to replace the code in correctPIN() and incorrectPIN() with the required code to run when a correct or incorrect sequence is entered.

Looking Ahead

Once again, you have learned another way to gather input for your Arduino. You've also gained the foundational knowledge to create a useful method of controlling a sketch using a numeric keypad, as well as the foundations for a combination lock to access anything that your Arduino can control. You've also learned the very useful switch-case function. Moving on to the next chapter you'll learn about another form of input: the touchscreen.

10

ACCEPTING USER INPUT
WITH TOUCHSCREENS

In this chapter you will

- Learn how to connect a resistive touchscreen to your Arduino
- Discover the values that can be returned from the touchscreen
- Create a simple on/off touch switch
- Create an on/off touch switch with a dimmer-style control

We see touchscreens everywhere today: smartphones, tablets, and portable video-game systems. So why not use a touchscreen to accept input from a user?

Touchscreens

Touchscreens can be quite expensive, but we'll use an inexpensive model available from SparkFun (part numbers LCD-08977 and BOB-09170), originally designed for the Nintendo DS game console.

This touchscreen, which measures about 2 by 2 3/4 inches, is shown mounted on a breadboard in Figure 10-1.

Figure 10-1: Touchscreen mounted on a solderless breadboard

Notice the horizontal ribbon cable connected into the small circuit board at the upper-right corner (it's circled in the figure). This *breakout board* is used to attach the Arduino and the breadboard to the touchscreen; Figure 10-2 shows a close-up of the board.

Figure 10-2: Touchscreen connector board

Connecting the Touchscreen

Connect the touchscreen breakout board to an Arduino as shown in Table 10-1.

Table 10-1: Touchscreen breakout board connections

Breakout Board Pin	Arduino Pin
Y1	A0
X1	A1
Y2	A2
X2	A3

Project #34: Addressing Areas on the Touchscreen

The touchscreen has two layers of resistive coating between the top layer of plastic film and the bottom layer of glass. One coating acts as the x-axis, and the other is the y-axis. As current passes through each coating, the resistance of the coating varies depending on where it has been touched, so when each current is measured, the X and Y positions of the touched area can be determined.

In this project, we'll use the Arduino to record touched locations on the screen and to convert the touches to integers that can be used to reference areas of the screen.

The Hardware

The following hardware is required:

- Touchscreen and breakout board
- One 10 kΩ trimpot
- One 16×2-character LCD module
- Various connecting wires
- One breadboard
- Arduino and USB cable

Connect the touchscreen as described in Table 10-1, and connect the LCD module as described in Figure 7-2 on page 149.

The Sketch

Enter and upload the following sketch. I've highlighted important aspects of the sketch with comments:

```
// Project 34 - Addressing Areas on the Touchscreen

#include <LiquidCrystal.h>
LiquidCrystal lcd(4,5,6,7,8,9);

int x,y = 0;

void setup()
{
  lcd.begin(16,2);
  lcd.clear();
}
```
❶ `int readX()   // returns the value of the touchscreen's x-axis`
```
  {
    int xr=0;
    pinMode(A0, INPUT);
    pinMode(A1, OUTPUT);
    pinMode(A2, INPUT);
```

```
    pinMode(A3, OUTPUT);
    digitalWrite(A1, LOW);    // set A1 to GND
    digitalWrite(A3, HIGH);   // set A3 as 5V
    delay(5);
    xr=analogRead(0);         // store the value of the x-axis
    return xr;
  }

❷ int readY()      // returns the value of the touchscreen's y-axis
  {
    int yr=0;
    pinMode(A0, OUTPUT);    // A0
    pinMode(A1, INPUT);     // A1
    pinMode(A2, OUTPUT);    // A2
    pinMode(A3, INPUT);     // A3
    digitalWrite(14, LOW);  // set A0 to GND
    digitalWrite(16, HIGH); // set A2 as 5V
    delay(5);
    yr=analogRead(1); // store the value of the y-axis
    return yr;
  }

  void loop()
  {
    lcd.setCursor(0,0);
❸  lcd.print(" x = ");
    x=readX();
    lcd.print(x);
    y=readY();
    lcd.setCursor(0,1);
❹  lcd.print(" y = ");
    lcd.print(y);
    delay (200);
  }
```

The functions readX() and readY() at ❶ and ❷ read the currents from the touchscreen's resistive layers, measure the current using analogRead(), and return the read value. The sketch rapidly runs these two functions to provide a real-time position of the screen area being touched and displays this on the LCD at ❸ and ❹. (The delay(5) in each function is required to allow the input/output pins time to change their states.)

Testing the Sketch

To test the sketch, monitor the LCD module while you touch the screen, and notice how the X and Y values change relative to the touched position on the screen. Also notice the values displayed when the screen is not being touched, such as those shown in Figure 10-3.

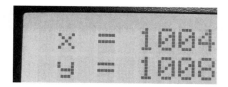

Figure 10-3: Values that appear when the touchscreen is not touched

It is important that you note these values, because you can use them to detect when the screen is not being touched in your sketch.

Mapping the Touchscreen

By touching the corners of the touchscreen and recording the values returned, you can actually map the touchscreen, as shown in Figure 10-4. Basically, you're plotting the coordinates for each corner. Once you have determined these values, you can divide up the touchscreen map into smaller areas for use as control surfaces.

Figure 10-4: A touchscreen map

After you've created your touchscreen map, you can divide it into smaller regions and use the functions readX() and readY() to create a variety of control regions on the screen, which you can then use with if-then statements to cause specific actions to occur depending on where the screen is touched, as you'll see in Project 35.

Project #35: Creating a Two-Zone On/Off Touch Switch

In this project, we'll use our touchscreen map to create an on/off switch. Start by dividing the touchscreen in half vertically, as shown in Figure 10-5: the left side will be "on" and the right side will be "off."

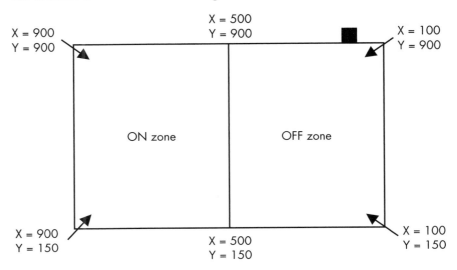

Figure 10-5: On/off switch map

The Arduino will determine which zone of the touchscreen is touched by comparing the recorded coordinates of the touches to the boundaries of each half of the screen. After the zone has been determined, a digital output could send the on or off signal, but for this sketch we'll simply display in the Serial Monitor whether a zone has been turned on or off.

The Sketch

Enter and upload the following sketch:

```
// Project 35 - Creating a Two-Zone On/Off Touch Switch

int x,y = 0;

void setup()
{
  Serial.begin(9600);
  pinMode(10, OUTPUT);
}

void switchOn()
{
  digitalWrite(10, HIGH);
  Serial.print("Turned ON at X = ");
  Serial.print(x);
```

```
    Serial.print(" Y = ");
    Serial.println(y);
    delay(200);
  }

  void switchOff()
  {
    digitalWrite(10, LOW);
    Serial.print("Turned OFF at X = ");
    Serial.print(x);
    Serial.print(" Y = ");
    Serial.println(y);
    delay(200);
  }

  int readX()                    // returns the value of the touchscreen's x-axis
  {
    int xr=0;
    pinMode(A0, INPUT);
    pinMode(A1, OUTPUT);
    pinMode(A2, INPUT);
    pinMode(A3, OUTPUT);
    digitalWrite(A1, LOW);     // set A1 to GND
    digitalWrite(A3, HIGH);    // set A3 as 5V
    delay(5);
    xr=analogRead(0);
    return xr;
  }

  int readY()                    // returns the value of the touchscreen's y-axis
  {
    int yr=0;
    pinMode(A0, OUTPUT);
    pinMode(A1, INPUT);
    pinMode(A2, OUTPUT);
    pinMode(A3, INPUT);
    digitalWrite(A0, LOW);     // set A0 to GND
    digitalWrite(A2, HIGH);    // set A2 as 5V
    delay(5);
    yr=analogRead(1);
    return yr;
  }

  void loop()
  {
    x=readX();
    y=readY();
```

❶ `// test for ON`

```
    if (x<=900 && x>=500)
    {
      switchOn();
    }
```

❷ // test for OFF

```
  if (x<500 && x>=100)
  {
    switchOff();
  }
}
```

How It Works

The two if functions used in void loop() check for a touch on the left or right side of the screen. If the left side is touched, the touch is detected as an "on" press at ❶. If the right side is touched (an "off" press), the touch is detected at ❷.

NOTE   *The y-axis is ignored because the touchscreen is split vertically. If we were to create horizontal boundaries, the y-axis would need to be checked as well—as you'll see in Project 36.*

Testing the Sketch

The resulting output of the sketch that appears in the Serial Monitor is shown in Figure 10-6. The status of the switch and the coordinates are shown after each screen touch.

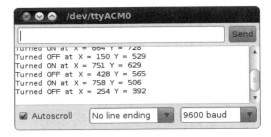

Figure 10-6: Output from Project 35

Project #36: Creating a Three-Zone Touch Switch

In this project we'll create a three-zone touch switch for an LED on digital pin 3 that turns the LED on or off and adjusts the brightness using PWM (Chapter 3).

The Touchscreen Map

Our touchscreen map is shown in Figure 10-7.

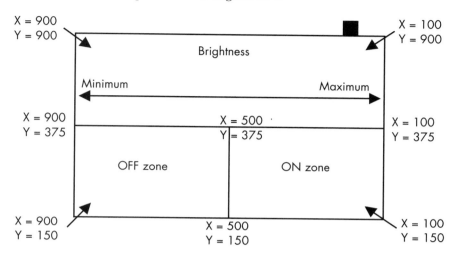

Figure 10-7: Touchscreen map for Project 36

The touchscreen map is divided into "off" and "on" zones and a "brightness" control zone. We measure the values returned by the touchscreen to determine which part has been touched and then act accordingly.

The Sketch

Enter and upload the following sketch:

```
// Project 36 - Creating a Three-Zone Touch Switch

int x,y = 0;

void setup()
{
  pinMode(3, OUTPUT);
}

void switchOn()
{
  digitalWrite(3, HIGH);
  delay(200);
}

void switchOff()
{
  digitalWrite(3, LOW);
  delay(200);
}
```

```
void setBrightness()
{
  int xx, bright;
  float br;
  xx=x-100;
❶ br=(800-xx)/255;
  bright=int(br);
  analogWrite(3, bright);
}

int readX()                    // returns the value of the touchscreen's x-axis
{
  int xr=0;
  pinMode(A0, INPUT);
  pinMode(A1, OUTPUT);
  pinMode(A2, INPUT);
  pinMode(A3, OUTPUT);
  digitalWrite(A1, LOW);   // set A1 to GND
  digitalWrite(A3, HIGH);  // set A3 as 5V
  delay(5);
  xr=analogRead(0);
  return xr;
}

int readY()                    // returns the value of the touchscreen's y-axis
{
  int yr=0;
  pinMode(A0, OUTPUT);       // A0
  pinMode(A1, INPUT);        // A1
  pinMode(A2, OUTPUT);       // A2
  pinMode(A3, INPUT);        // A3
  digitalWrite(A0, LOW);    // set A0 to GND
  digitalWrite(A2, HIGH);   // set A2 as 5V
  delay(5);
  yr=analogRead(1);
  return yr;
}

void loop()
{
  x=readX();
  y=readY();

// test for ON

❷ if (x<=500 && x>=100 && y>= 150 && y<375)
    {
      switchOn();
    }
```

```
   // test for OFF

❸ if (x>500 && x<=900 && y>= 150 && y<375)
   {
     switchOff();
   }

   // test for brightness

❹ if (y>=375 && y<=900)
   {
     setBrightness();
   }
}
```

How It Works

As with the sketch for the two-zone map, this sketch will check for touches in the "on" and "off" zones (which are now smaller) at ❷ and ❸ and for any touches above the horizontal divider, which we'll use to determine brightness at ❹. If the screen is touched in the brightness area, the position on the x-axis is converted to a relative value for PWM at ❶ and the LED is adjusted accordingly using the function setBrightness().

You can use these same basic functions to create any number of switches or sliders with this simple and inexpensive touchscreen.

Looking Ahead

This chapter introduced you to another method of accepting user data and controlling your Arduino. In the next chapter we'll focus on the Arduino board itself, learn about the different versions available, and create our own version on a solderless breadboard.

11

MEET THE ARDUINO FAMILY

In this chapter you will

- Learn how to build your own Arduino circuit on a solderless breadboard
- Explore the features and benefits of a wide range of Arduino-compatible boards
- Learn about open-source hardware

We'll break down the Arduino design into a group of parts, and then you'll build your own Arduino circuit on a solderless breadboard. Building your own circuit can save you money, especially when you're working with changing projects and prototypes. You'll also learn about some new components and circuitry. Then we'll explore ways to upload sketches to your homemade Arduino that don't require extra hardware. Finally, we'll examine the more common alternatives to the Arduino Uno and explore their differences.

Project #37: Creating Your Own Breadboard Arduino

As your projects and experiments increase in complexity or number, the cost of purchasing Arduino boards for each task can easily get out of hand, especially if you like to work on more than one project at a time. At this point, it's cheaper and easier to integrate the circuitry of an Arduino board into your project by building an Arduino circuit on a solderless breadboard that you can then expand for your specific project. It should cost less than $10 in parts to reproduce the basic Arduino circuitry on a breadboard (which itself is usually reusable if you're not too hard on it). It's easier to make your own if your project has a lot of external circuitry, because it saves you running lots of wires from an Arduino back to the breadboard.

The Hardware

To build a minimalist Arduino, you'll need the following hardware:

- One breadboard
- Various connecting wires
- One 7805 linear voltage regulator
- One 16 MHz crystal oscillator (such as Newark part number 16C8140)
- One ATmega328P-PU microcontroller with Arduino bootloader
- One 1 µF, 25 V electrolytic capacitor (C1)
- One 100 µF, 25 V electrolytic capacitor (C2)
- Two 22 pF, 50 V ceramic capacitors (C3 and C4)
- Two 100 nF, 50 V ceramic capacitors (C5)
- Two 560 Ω resistors (R1 and R2)
- One 10 kΩ resistor (R3)
- Two LEDs of your choice (LED1 and LED2)
- One push button (S1)
- One six-way header pin
- One PP3-type battery snap
- One 9 V PP3-type battery

Some of these parts might be new to you. In the following sections, we'll explain each part and show you an example and a schematic of each.

7805 Linear Voltage Regulator

A *linear voltage regulator* contains a simple circuit that converts one voltage to another. The regulator included in the parts list is the 7805-type, which can convert a voltage between 7 and 30 volts to a fixed 5 volts, with a current up to 1 amp, which is perfect for running our breadboard Arduino. Figure 11-1 shows an example of a 7805 called a TO-220 style, which refers to its physical shape.

Figure 11-1: 7805 linear voltage regulator

Figure 11-2 shows the schematic symbol for the 7805. When you're looking at the labeled side of the 7805, the pin on the left is for input voltage, the center pin connects to GND, and the right-hand pin is the 5 V output connection. The metal tab at the top is drilled to allow it to connect to a larger piece of metal known as a *heat sink*. We use a heat sink when the circuit draws up to the maximum of 1 amp of current, because the 7805 will become quite warm at that level of use. The metal tab is also connected to GND. We will need one 7805 regulator for our example.

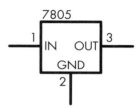

Figure 11-2: 7805 schematic symbol

16 MHz Crystal Oscillator

More commonly known as simply a *crystal*, the crystal oscillator creates an electrical signal with a very accurate frequency. In this case, the frequency is 16 MHz. The crystal we'll use is shown in Figure 11-3.

Figure 11-3: Crystal oscillator

Compare this image to the crystal on your Arduino board. They should be identical in shape and size.

Crystals are not polarized. The schematic symbol is shown in Figure 11-4.

The crystal determines the microcontroller's speed of operation. For example, the microcontroller circuit we'll be assembling runs at 16 MHz, which means it can execute 16 million processor instructions per second. That doesn't mean it can execute a line of a sketch or a function that rapidly, however, since it takes many processor instructions to interpret a single line of code.

Figure 11-4: Crystal oscillator schematic symbol

Atmel ATmega328P-PU Microcontroller IC

A *microcontroller* is a tiny computer that contains a processor that executes instructions, various types of memory to hold data and instructions from our sketch, and various ways to send and receive data. As explained in Chapter 2, the microcontroller is the brains of our breadboard Arduino. An example of the ATmega328P-PU is shown in Figure 11-5. When looking at the IC in the photo, notice that pin number 1 is at the bottom-left of the IC and is marked by a small dot.

Figure 11-5: ATmega328P-PU

The schematic symbol for the microcontroller is shown in Figure 11-6.

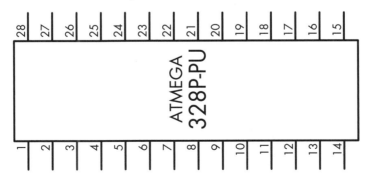

Figure 11-6: Microcontroller schematic symbol

Microcontrollers don't all contain the Arduino *bootloader*, the software that allows it to interpret sketches written for an Arduino. When choosing a microcontroller to include in a homemade Arduino, be sure to select one that already includes the bootloader. These are generally available from the same retailers that sell Arduino boards, such as adafruit, Freetronics, and SparkFun.

The Schematic

Figure 11-7 shows the circuit schematic.

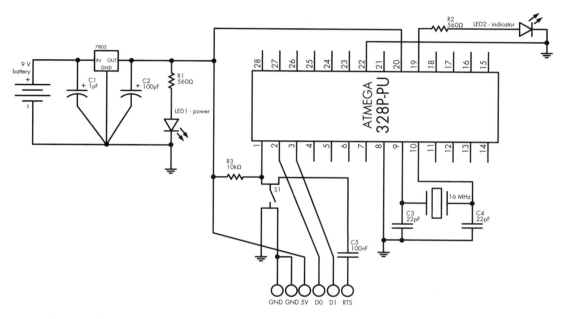

Figure 11-7: Breadboard Arduino schematic

The schematic contains two sections. The first, on the left, is the power supply, which reduces the voltage to a smooth 5 V. You'll see an LED that is lit when the power is on. The second section, on the right, consists of the microcontroller, the reset button, the programming pins, and another LED. This LED is wired to the ATmega328P-PU pin that is used as Arduino pin 13. Use the schematic to wire up your Arduino. Don't forget to run the wires to the six-way header pin (shown in Figure 11-8), represented by the six circles at the bottom of the schematic. We'll use this connection later in the chapter to upload a sketch to our homemade Arduino.

The circuit will be powered using a simple 9 V battery and matching snap connector, as shown in Figure 11-9. Connect the red lead of the battery snap connector to the positive (+) point and the black lead to the negative (−) point on the left side of the circuit.

Figure 11-8: Six-way header pin

Figure 11-9: 9 V battery and snap connector

Identifying the Arduino Pins

Where are all the Arduino pins on our homemade Arduino? All the analog, digital, and other pins available on the normal Arduino board are also available in our breadboard version; you simply need to connect directly to the microcontroller.

In our breadboard Arduino, the R2 and LED2 are on digital pin 13. Table 11-1 lists the Arduino pins on the left and the matching ATmega328P-PU pins on the right.

Table 11-1: Pins for ATmega328P-PU

Arduino Pin Name	ATmega328P-PU Pin
RST	1
RX/D0	2
TX/D1	3
D2	4
D3	5
D4	6
(5 V only)	7
GND	8
D5	11
D6	12
D7	13
D8	14
D9	15
D10	16
D11	17
D12	18
D13	19
(5 V only)	20
AREF	21
GND	22
A0	23
A1	24
A2	25
A3	26
A4	27
A5	28

To avoid confusion, retailers such as adafruit and Freetronics sell adhesive labels to place over the microcontroller, like the one shown on the microcontroller in Figure 11-10 (order at *http://www.freetronics.com/ mculabel/*).

Figure 11-10: Pin labels

Running a Test Sketch

Now it's time to upload a sketch. We'll start by uploading a simple sketch to blink the LED:

```
// Project 37 - Creating Your Own Breadboard Arduino

void setup()
{
  pinMode(13, OUTPUT);
}

void loop()
{
  digitalWrite(13, HIGH);
  delay(1000);
  digitalWrite(13, LOW);
  delay(1000);
}
```

You can upload the sketch in one of three ways.

Use the Microcontroller Swap Method

The most inexpensive way to upload a sketch is to remove the microcontroller from an existing Arduino, insert the microcontroller from your homemade Arduino, upload the sketch, and then swap the microcontrollers again.

To remove a microcontroller from the Arduino safely, use an IC extractor, as shown in Figure 11-11.

Figure 11-11: Use an IC extractor to remove a microcontroller.

When removing the microcontroller, be sure to pull both ends out evenly and slowly *at the same time*, and take your time! Removing the component might be difficult, but eventually the microcontroller will come out.

When inserting a microcontroller into the breadboard or your Arduino, you may have to bend the pins a little to make them perpendicular with the body of the microcontroller so that they can slide in easily. To do this, place one side of the component against a flat surface and gently push down; then repeat on the other side, as shown in Figure 11-12.

Finally, when you return the original microcontroller to your Arduino board, remember that the end with the notch should be on the right side, as shown in Figure 11-13.

Figure 11-12: Bending the microcontroller pins

Figure 11-13: Correct orientation of the microcontroller in an Arduino

Connect to an Existing Arduino Board

We can use the USB interface of an Arduino Uno to upload sketches to the microcontroller in our breadboard Arduino. Using this method reduces wear on the Arduino board's socket and saves you money, because you won't need to buy a separate USB programming cable.

Here's how to upload a sketch to the microcontroller using the USB interface:

1. Remove the microcontroller from your Arduino Uno, and unplug the USB cable.
2. Remove the power (if connected) from the breadboard Arduino circuit.
3. Connect a wire from Arduino digital pin 0 to pin 2 of the breadboard's ATmega328P-PU, and connect another from Arduino digital pin 1 to pin 3 of the ATmega328P-PU.
4. Connect the 5 V and GND from the Uno to the matching areas on the breadboard.
5. Connect a wire from Arduino RST to pin 1 of the ATmega328P-PU.
6. Plug the USB cable into the Arduino Uno board.

At this point, the system should behave as if it were an ordinary Arduino Uno, so you should be able to upload sketches into the breadboard circuit's microcontroller normally and use the serial monitor if necessary.

Use an FTDI Programming Cable

The final method is the easiest, but it requires the purchase of a USB programming cable, known as an *FTDI cable* (simply because the USB interface circuitry inside is made by a company called FTDI). When purchasing an FTDI cable, make sure it's the 5 V model, because the 3.3 V model will not work properly. This cable (shown in Figure 11-14) has a USB plug on one end and a socket with six wires on the other. The USB end of this cable contains circuitry equivalent to the USB interface on an Arduino Uno board. The six-wire socket connects to the header pins shown in Figures 11-7 and 11-8.

Figure 11-14: FTDI cable

When you're connecting the cable, be sure that the side of the socket with the black wire connects to the GND pin on the breadboard's header pins. Once the cable is connected, it also supplies power to the circuit, just like a normal Arduino board would do.

Before uploading your sketch or using the serial monitor, change the board type to Arduino Duemilanove or Nano w/ ATmega328 by choosing **Tools ▸ Board** and then selecting the correct microcontroller (Figure 11-15).

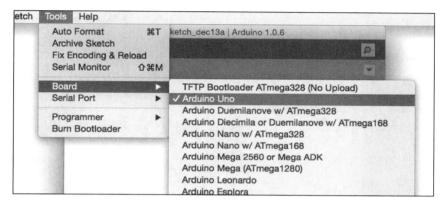

Figure 11-15: Changing the board type in the IDE

Once you have selected a method of uploading, test it by uploading the Project 37 sketch. Now you should be able to design more complex circuits using only a breadboard, which will let you create more projects for less money. You can even build more permanent projects from scratch if you learn to make your own printed circuit boards.

The Many Arduino Boards

Although we have been working exclusively with the Arduino Uno board throughout the book, you can choose from many alternative boards. These will vary in physical size, the number of input and output pins, memory space for sketches, and purchase price.

One of the crucial differences between boards is the microcontroller used. Current boards generally use the ATmega328 or the ATmega2560 microcontroller, and the Due uses another, more powerful version. The main differences between the two (including both versions of the ATmega328) are summarized in Table 11-2.

Table 11-2: Microcontroller Comparison Chart

	ATmega328P-PU	ATmega328P SMD	ATmega2560	SAM3X8E
User replaceable?	Yes	No	No	No
Processing speed	16 MHz	16 MHz	16 MHz	84 MHz
Operating voltage	5 V	5 V	5 V	3.3 V
Number of digital pins	14 (6 PWM–capable)	14 (6 PWM–capable)	54 (14 PWM–capable)	54 (12 PWM–capable)
Number of analog input pins	6	8	16	12
DC current per I/O pin	40 mA	40 mA	40 mA	3 to 15 mA
Available flash memory	31.5KB	31.5KB	248KB	512KB
EEPROM size	1KB	1KB	4KB	No EEPROM
SRAM size	2KB	2KB	8KB	96KB

The main parameters used to compare various Arduino-compatible boards are the types of memory they contain and the amount of each type. Following are the three types of memory:

- *Flash memory* is the space available to store a sketch after it has been compiled and uploaded by the IDE.
- *EEPROM (electrically erasable programmable read-only memory)* is a small space that can store byte variables, as you'll learn in Chapter 16.
- *SRAM* is the space available to store variables from your programs.

NOTE *Many Arduino boards are available in addition to the Uno, and the few described here are only the tip of the iceberg. When you're planning large or complex projects, don't be afraid to scale up to the larger Mega boards. By the same token, if you need only a few I/O pins for a more permanent project, consider the Nano or even a LilyPad.*

Let's explore the range of the available boards.

Arduino Uno

The Uno is currently considered the standard Arduino board. All Arduino shields ever made should be compatible with the Uno. The Uno is considered to be the easiest-to-use Arduino board due to its built-in USB interface and removable microcontroller.

Freetronics Eleven

Many boards on the market emulate the function of the Arduino Uno, and some have even improved on the standard design. One of these is the Freetronics Eleven, shown in Figure 11-16.

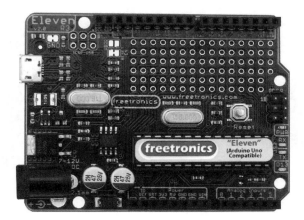

Figure 11-16: Freetronics Eleven

Although the Eleven is completely compatible with the Arduino Uno, it offers several improvements that make it a worthwhile product. The first is the large prototyping area just below the digital I/O pins. This area allows you to construct your own circuit directly on the main board, which can save space and money, since you won't need to purchase a separate prototyping shield.

Second, the transmitter/receiver (TX/RX), power, and D13 LEDs are positioned on the far-right side of the board; this placement allows them to be visible even when a shield is attached. Finally, it uses a micro-USB socket, which is much smaller than the standard USB socket used on the Uno. This makes designing your own shield simpler, since you don't have to worry about your connections bumping into the USB socket. It is available from *http://www.freetronics.com/products/eleven/*.

The Freeduino

The Freeduino board comes from a collaborative open source project that describes and publishes files to allow people to construct their own Arduino-compatible boards. One of the more popular designs is the Duemilanove-compatible board kit, shown in Figure 11-17.

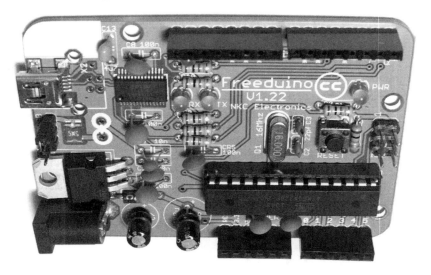

Figure 11-17: An assembled Duemilanove-compatible Freeduino board

You can use the Freeduino board to run all the projects in this book. Two of the major benefits of the Freeduino are that it's cheap and that assembling a board by hand is intrinsically satisfying. The Freeduino kit is available from *http://www.seeedstudio.com/*.

The Pro Trinket

The Pro Trinket (Figure 11-18) is a miniaturized version of the Arduino Uno designed for working with solderless breadboards, wearable electronics, or any situation where you need a much smaller board.

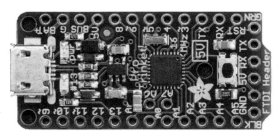

Figure 11-18: Adafruit Pro Trinket

There are some slight differences from the Arduino Uno (for example, no serial output unless you use an external FTDI cable); however, for the price this is a great value board. The Pro Trinket is available from *http://www.adafruit.com/*.

The Arduino Nano

When you need a compact, assembled Arduino-compatible board, the Nano should fit the bill. Designed to work in a solderless breadboard, the Nano (Figure 11-19) is a tiny but powerful Arduino.

Figure 11-19: An Arduino Nano

The Nano measures only 0.7 inches by 1.7 inches, yet it offers all the functionality of the Freeduino. Furthermore, it uses the SMD version of the ATmega328P, so it has two extra analog input pins (A6 and A7). The Nano is available from *http://www.gravitech.us/arna30wiatp.html*.

The Arduino LilyPad

The LilyPad is designed to be integrated inside creative projects, such as wearable electronics. In fact, you can actually wash a LilyPad with water and a mild detergent, so it's ideal to use for lighting up a sweatshirt, for example. The board design is unique, as shown in Figure 11-20.

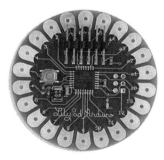

Figure 11-20: An Arduino LilyPad

The I/O pins on the LilyPad require that wires be soldered to the board, so the LilyPad is more suited for use with permanent projects. As part of its minimalist design, it has no voltage regulation circuitry, so it's up to the user to provide his or her own supply between

2.7 and 5.5 V. The LilyPad also lacks a USB interface, so a 5 V FTDI cable is required to upload sketches. You can get Arduino LilyPad boards from almost any Arduino retailer.

The Arduino Mega 2560

When you run out of I/O pins on your Arduino Uno or you need space for much larger sketches, consider a Mega 2560. It is physically a much larger board than the Arduino, measuring 4.3 inches by 2.1 inches; it's shown in Figure 11-21.

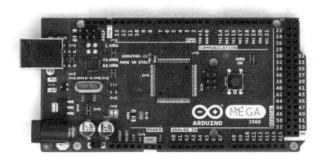

Figure 11-21: Arduino Mega 2560

Although the Mega 2560 board is much larger than the Uno, you can still use most Arduino shields with it, and Mega-sized prototyping shields are available for larger projects that the Uno can't accommodate. Since the Mega uses the ATmega2560 microcontroller, its memory space and I/O capabilities (as described in Table 11-2) are much greater than those of the Uno. Additionally, four separate serial communication lines increase its data transmission capabilities. You can get Mega 2560 boards from almost any Arduino retailer.

The Freetronics EtherMega

When you need an Arduino Mega 2560, a microSD card shield, and an Ethernet shield to connect to the Internet, your best alternative is an EtherMega (Figure 11-22), because it has all these functions on a single board and is less expensive than purchasing each component separately. The EtherMega is available from *http://www.freetronics.com/ethermega/*.

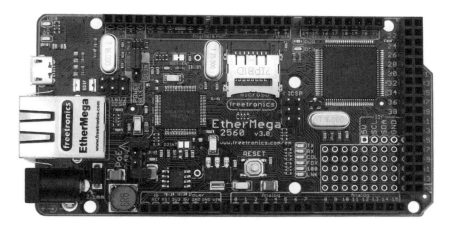

Figure 11-22: Freetronics EtherMega

The Arduino Due

With an 84 MHz processor that can run your sketches much faster, this is the most powerful Arduino board ever released. As you can see in Figure 11-23, the board is quite similar to the Arduino Mega 2560, but there is an extra USB port for external devices and different pin labels.

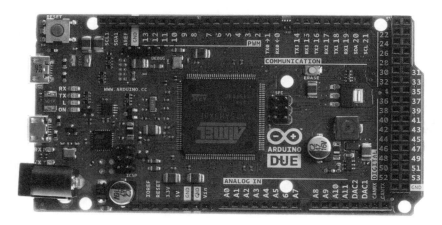

Figure 11-23: Arduino Due

Furthermore, the Due has just over 16 times the memory of an Uno board, so you can really create complex and detailed sketches. However, the Due operates only on 3.3 V—so any circuits, shields, or other devices connected to the analog or digital pins cannot have a voltage greater than 3.3 V. Despite these limitations, the benefits of using the Due outweigh the changes in the hardware.

NOTE *When shopping for your next Arduino board or accessory, be sure to buy from a reputable retailer that offers support and a guarantee. Although the Internet is flooded with cheap alternatives, corners are often cut to produce products at abnormally low prices, and you might have no way of seeking recompense if you're sold a faulty or incorrectly specified product.*

OPEN SOURCE HARDWARE

The Arduino hardware design is released to the public so that anyone can manufacture, modify, distribute, and use it as they see fit. This type of distribution falls under the umbrella of *open source hardware*—a recent movement that is an antithesis to the concept of copyrights and legal protection of intellectual property. The Arduino team decided to allow its designs to be free for the benefit of the larger hardware community and for the greater good.

In the spirit of open source hardware, many organizations that produce accessories or modifications of the original Arduino boards publish their designs under the same license. This allows for a much faster process of product improvement than would be possible for a single organization developing the product alone.

Looking Ahead

This chapter has given you a broader picture of the types of hardware available, including a breadboard Arduino that you build yourself. You've seen the parts that make up the Arduino design, and you've seen how to build your own Arduino using a solderless breadboard. You now know how to make more than one Arduino-based prototype without having to purchase more boards. You also know about the variety of Arduino boards on the market, and you should be able to select the Arduino board that best meets your needs. Finally, you've gained an understanding of the Arduino open source movement itself.

In the next chapter you'll learn to use a variety of motors and begin working on your own Arduino-controlled motorized tank!

12

MOTORS AND MOVEMENT

In this chapter you will

- Use a servo to create an analog thermometer
- Learn how to control the speed and direction of electric motors
- Use an Arduino motor shield
- Begin work on a motorized tank robot
- Use simple microswitches for collision avoidance
- Use infrared and ultrasonic distance sensors for collision avoidance

Making Small Motions with Servos

A *servo* (short for *servomechanism*) contains an electric motor that can be commanded to rotate to a specific angular position. For example, you might use a servo to control the steering of a remote control car by connecting the servo to a *horn*, a small arm or bar that the servo rotates. An example of a horn is one of the hands on an analog clock. Figure 12-1 shows a servo and three types of horns.

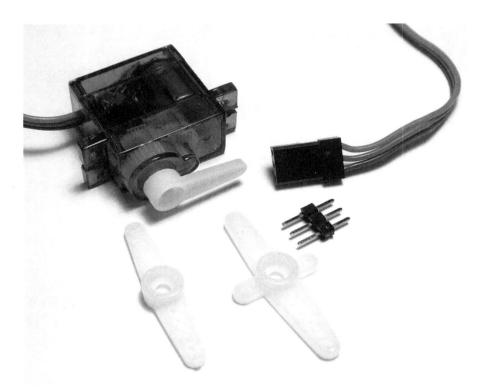

Figure 12-1: Servo and various horns

Selecting a Servo

When you're selecting a servo, consider several parameters:

- **Speed**　The time it takes for the servo to rotate, usually measured in seconds per angular degree.
- **Rotational range**　The angular range through which the servo can rotate—for example, 180 degrees (half of a full rotation) or 360 degrees (one complete rotation).
- **Current**　How much current the servo draws. When using a servo with an Arduino, you may need to use an external power supply for the servo.
- **Torque**　The amount of force the servo can exert when rotating. The greater the torque, the heavier the item the servo can control. The torque produced is generally proportional to the amount of current used.

The servo shown in Figure 12-1 is a hexTronik HXT900. It is inexpensive and can rotate up to 180 degrees, as shown in Figure 12-2.

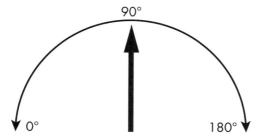

Figure 12-2: Example servo rotation range

Connecting a Servo

It's easy to connect a servo to an Arduino because only three wires are involved. If you're using the HXT900, the darkest wire connects to GND, the center wire connects to 5 V, and the lightest wire (the *pulse* wire) connects to a digital pin. If you're using a different servo, check its data sheet for the correct wiring.

Putting a Servo to Work

Now let's put our servo to work. In this sketch, the servo will turn through its rotational range. Connect the servo to your Arduino as described, with the pulse wire connected to digital pin 4, and then enter and upload the sketch in Listing 12-1.

```
// Listing 12-1

#include <Servo.h>
Servo myservo;

void setup()
{
  myservo.attach(4);
}

void loop()
{
    myservo.write(180);
    delay(1000);
    myservo.write(90);
    delay(1000);
    myservo.write(0);
    delay(1000);
}
```

Listing 12-1: Servo demonstration sketch

In this sketch, we use the *servo* library included with the Arduino IDE and create an instance of the servo with the following:

```
#include <Servo.h>
Servo myservo;
```

Then, in `void setup()`, we tell the Arduino which digital pin the servo control is using:

```
myservo.attach(4); // control pin on digital four
```

Now we simply move the servo with the following:

```
myservo.write(x);
```

Here, x is an integer between 0 and 180 degrees—the angular position to which the servo will be moved. When running the sketch in Listing 12-1, the servo will rotate across its maximum range, stopping at the extremes (0 degrees and 180 degrees) and at the midpoint (90 degrees). When looking at your servo, note that the 180-degree position is on the left and 0 degrees is on the right.

In addition to pushing or pulling objects, servos can also be used to communicate data similar to an analog gauge. For example, you could use a servo as an analog thermometer, as you'll see in Project 38.

Project #38: Building an Analog Thermometer

Using our servo and the TMP36 temperature sensor from earlier chapters, we'll build an analog thermometer. We'll measure the temperature and then convert this measurement to an angle between 0 and 180 degrees to indicate a temperature between 0 and 30 degrees Celsius. The servo will rotate to the angle that matches the current temperature.

The Hardware

The required hardware is minimal:

- One TMP36 temperature sensor
- One breadboard
- One small servo
- Various connecting wires
- Arduino and USB cable

The Schematic

The circuit is also very simple, as shown in Figure 12-3.

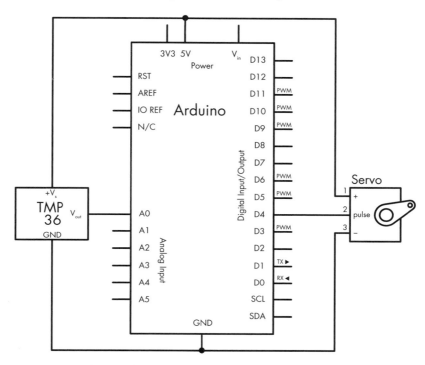

Figure 12-3: Schematic for Project 38

The Sketch

The sketch will determine the temperature using the same method used in Project 8. Then it will convert the temperature into an angular rotation value for the servo.

Enter and upload the following sketch:

```
// Project 38 - Building an Analog Thermometer

float voltage = 0;
float  sensor = 0;
float currentC = 0;
int      angle = 0;

#include <Servo.h>
Servo myservo;

void setup()
{
  myservo.attach(4);
}
```

```
int calculateservo(float temperature)
{
  float resulta;
  int resultb;
  resulta = -6 * temperature;
  resulta = resulta + 180;
  resultb = int(resulta);
  return resultb;
}

void loop()
{
  // read current temperature
  sensor = analogRead(0);
  voltage = (sensor*5000)/1024;
  voltage = voltage-500;
  currentC = voltage/10;

  // display current temperature on servo
  angle = calculateservo(currentC);
  // convert temperature to a servo position
  if (angle>=0 && angle <=30)
  {
    myservo.write(angle); // set servo to temperature
    delay(1000);
  }
}
```

Most of this sketch should be clear to you at this point, but the function calculateservo() is new. This function converts the temperature into the matching angle for the servo to use according to the following formula:

$$angle = (-6 \times temperature) + 180$$

You might find it useful to make a *backing sheet* to show the range of temperatures that the servo will display, with a small arrow to create a realistic effect. An example backing sheet is shown in Figure 12-4.

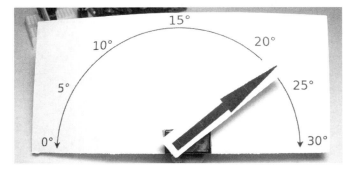

Figure 12-4: A backing sheet indicates the temperature on our thermometer.

Using Electric Motors

The next step in our motor-controlling journey is to work with small electric motors. Small motors are used for many applications, from small fans to toy cars to model railroads. As with servos, you need to consider several parameters when you're choosing an electric motor:

- **The operating voltage** This can vary, from 3 V to more than 12 V.
- **The current without a load** The amount of current the motor uses at its operating voltage while spinning freely, without anything connected to the motor's shaft.
- **The stall current** The amount of current used by the motor when it is trying to turn but cannot because of the load on the motor.
- **The speed at the operating voltage** The motor's speed in revolutions per minute (RPM).

Our example will use a small, inexpensive electric motor with a speed of 8,540 RPM when running on 3 V, similar to the one shown in Figure 12-5.

To control our motor, we'll use a transistor, which was described in Chapter 3. Because our motor uses up to 0.7 A of current (more than can be passed by the BC548 transistor), we'll use a transistor called a Darlington for this project.

Figure 12-5: Our small electric motor

The TIP120 Darlington Transistor

A *Darlington transistor* can handle high currents and voltages. The TIP120 Darlington can pass up to 5 A of current at 60 V, which is more than enough to control our small motor. The TIP120 uses a similar schematic symbol as the BC548, as shown in Figure 12-6, but the TIP120 transistor is physically larger than the BC548.

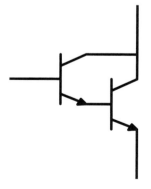

Figure 12-6: TIP120 schematic symbol

The TIP120 uses the TO-220 housing style, as shown in Figure 12-7.

Figure 12-7: The TIP120

When you're looking at the TIP120 from the labeled side, the pins from left to right are base (B), collector (C), and emitter (E). The metal heat sink tab is also connected to the collector.

Project #39: Controlling the Motor

In this project, we'll control the motor by adjusting the speed.

The Hardware

The following hardware is required:

- One small 3 V electric motor
- One 1 kΩ resistor (R1)
- One breadboard
- One 1N4004 diode
- One TIP120 Darlington transistor
- A separate 3 V power source
- Various connecting wires
- Arduino and USB cable

When working with motors, you must use a separate power source for them, because the Arduino cannot supply enough current for the motor in all possible situations. If the motor becomes stuck, then it will draw up to its *stall current*, which could be more than 1 amp. That's more than the Arduino can supply, and if it attempts to supply that much current, the Arduino could be permanently damaged.

A separate battery pack is a simple solution. For a 3 V supply, a two-cell AA battery pack with flying leads will suffice, such as the one shown in Figure 12-8.

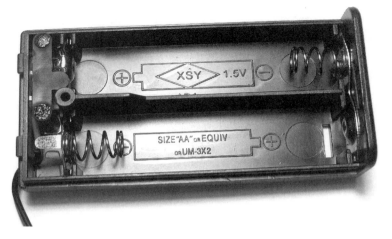

Figure 12-8: Two-cell AA battery pack

The Schematic

Assemble the circuit as shown in the schematic in Figure 12-9.

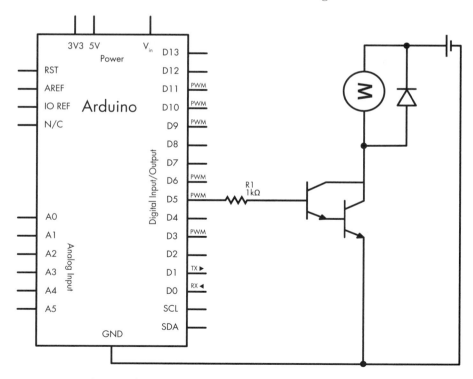

Figure 12-9: Schematic for Project 39

The Sketch

In this project, we'll adjust the speed of the motor from still (zero) to the maximum and then reduce it back to zero. Enter and upload the following sketch:

```
// Project 39 - Controlling the Motor

void setup()
{
  pinMode(5, OUTPUT);
}

void loop()
{
❶  for (int a=0; a<256; a++)
  {
    analogWrite(5, a);
❷    delay(100);
  }
❸  delay(5000);
❹  for (int a=255; a>=0; a--)
  {
    analogWrite(5,a);
    delay(100);
  }
  delay(5000);
}
```

We control the speed of the motor using pulse-width modulation (as explained in Project 3). Recall that we can do this only with digital pins 3, 5, 6, 9, 10, and 11. Using this method, current is applied to the motor in short bursts: the longer the burst, the faster the speed, as the motor is on more than it is off during a set period of time. So at ❶, the motor speed starts from still and increases slowly; you can control the acceleration by changing the delay value at ❷. At ❸, the motor is running as fast as possible and holds that speed for 5 seconds. Then, from ❹, the process reverses and the motor slows to a stop.

NOTE   *When it starts moving, you may hear a whine from the motor, which sounds similar to the sound of an electric train or a tram when it moves away from a station. This is normal and nothing to worry about.*

The diode is used in the same way it was with the relay control circuit described in Figure 3-19 on page 52 to protect the circuit. When the current is switched off from the motor, stray current exists for a brief amount of time inside the motor's coil and has to go somewhere. The diode allows the stray current to loop around through the coil until it dissipates as a tiny amount of heat.

Project #40: Building and Controlling a Tank Robot

Although controlling the speed of one motor can be useful, let's move into more interesting territory by controlling two motors at once to affect their speed *and* direction. Our goal is to describe the construction of a tank-style robot that we'll continue to work on in the next few chapters. Here we'll describe the construction and basic control of our tank.

Our tank has two motors that each control one tread, allowing it to climb over small obstacles, rotate in one position, and not crash into obstacles as it travels. You will be able to control the speed and direction of travel, and you will also learn how to add parts for collision avoidance and remote control. Once you have completed the projects in this book, you will have a solid foundation for creating your own versions and bringing your ideas to life.

The Hardware

The following hardware is required:

- One Pololu RP5 Tank Chassis package
- One Pololu RP5 Chassis plate
- Six alkaline AA cells
- One 9 V battery to DC socket cable
- A DFRobot 2A Arduino Motor Shield
- Arduino and USB cable

The Chassis

The foundation of any robot is a solid chassis containing the motors, drivetrain, and the power supply. An Arduino-powered robot also needs to have room to mount the Arduino and various external parts.

You can choose from many chassis models available on the market, but we'll use a tank chassis—the Pololu RP5 series shown in Figure 12-10, which contains two motors. You can also use the Pololu Daga Rover 5 tracked chassis.

Figure 12-10: Our tank chassis

Two Power Supplies

The Pololu chassis includes a holder for six AA cells, which we'll use as the power supply for the motors, as shown in Figure 12-11. The battery holder sits in the base of the chassis between the motors and gives the robot a low center of gravity.

Figure 12-11: Battery holder with six AA cells

Although the power supply in Figure 12-11 is large, we need to use a separate power supply for our Arduino board, because this power will allow the sketch to keep operating even if the motors fail. The power for the Arduino in this project comes from a 9 V battery, which can be connected to the power socket of the Arduino board using the cable shown in Figure 12-12.

Figure 12-12: Battery cable used to connect the battery to the Arduino

The Mounting Plate

The last part of our chassis is the *mounting plate*, which is shown in Figure 12-13.

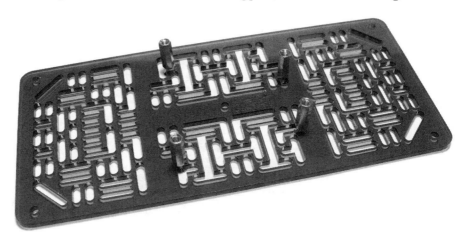

Figure 12-13: Mounting plate

The mounting plate covers the top of the chassis and allows you to bolt items on top using spacers and matching *M3 screws*. (Screws, spacers, and washers should be available from a robotics parts supplier or a large hardware store.) In Figure 12-14, you can see the mounting plate's spacers already fitted to hold our Arduino board.

Figure 12-14: Arduino mounted on the plate

The Schematic

The final requirement is to create the circuitry to control the two motors in the chassis. Although we could use the circuitry shown in Figure 12-9 for each of the motors, this wouldn't allow us to control the direction of the motor and could be somewhat inconvenient to wire up ourselves. Instead, we use a *motor shield*. A motor shield contains the circuitry we need to handle the higher current drawn by the motors and also accepts commands from the Arduino to control the speed and direction of both motors. For our tank, we'll use a 2A Motor Shield for Arduino from DFRobot (*http://www.dfrobot.com/*), as shown in Figure 12-15.

Figure 12-15: DFRobot motor shield

Connecting the Motor Shield

Making the required connections to the motor shield is simple: Connect the wires from the battery pack to the terminal block at the bottom-left of the shield, as shown in Figure 12-16. The black wire (negative) must be on the right side and the red wire on the left.

Next connect the two pairs of wires from the motors. Make sure the colors of the wires match the connections, as shown in Figure 12-17.

Figure 12-16: DC power connection

Figure 12-17: Connecting the motors

Connecting the Jumpers

The final task to set up the shield is to connect the appropriate jumpers. Look between the DC power connection and the bottom row of sockets on the shield, and you should see six pins with two black jumpers. Place them horizontally so that they cover the four pins on the left, as shown in Figure 12-18. Lastly, ensure that the four jumpers are connected vertically across the PWM jumpers, as shown in Figure 12-19.

Figure 12-18: Setting the correct power jumpers

Figure 12-19: Setting the correct mode jumpers

If your motor's wires are not color-coded, you may have to swap them after the first run to determine which way is forward or backward.

After you've connected the wiring and jumpers, inserted the battery pack, fitted the Arduino and shield to the mounting plate, and fastened it to the chassis, your tank should look something like the one in Figure 12-20.

Figure 12-20: Our tank bot is ready for action!

The Sketch

Now to get the tank moving. To begin, let's create some functions to simplify the movements. Because two motors are involved, we'll need four movements:

- Forward motion
- Reverse motion
- Rotate clockwise
- Rotate counterclockwise

Our motor shield controls each motor with two digital pins: One pin is for speed control using PWM (as demonstrated in Project 39), and the other determines the direction the motor will turn.

Four functions in our sketch match our four movements: goForward(), goBackward(), rotateLeft(), and rotateRight(). Each accepts a value in milliseconds, which is the length of time required to operate the movement, and a PWM speed value between 0 and 255. For example, to move forward for 2 seconds at full speed, we'd use goForward(2000,255).

Enter and save the following sketch (but don't upload it just yet):

```
// Project 40 - Building and Controlling a Tank Robot

int m1speed=6; // digital pins for speed control
int m2speed=5;
int m1direction=7; // digital pins for direction control
int m2direction=4;

void setup()
{
  pinMode(m1direction, OUTPUT);
  pinMode(m2direction, OUTPUT);
  delay(5000);
}

void goForward(int duration, int pwm)
{
  digitalWrite(m1direction,HIGH); // forward
  digitalWrite(m2direction,HIGH); // forward
  analogWrite(m1speed, pwm); // speed
  analogWrite(m2speed, pwm);
  delay(duration);
  analogWrite(m1speed, 0); // speed
  analogWrite(m2speed, 0);
}

void goBackward(int duration, int pwm)
{
  digitalWrite(m1direction,LOW); // backward
  digitalWrite(m2direction,LOW); // backward
  analogWrite(m1speed, pwm); // speed
  analogWrite(m2speed, pwm);
  delay(duration);
  analogWrite(m1speed, 0); // speed
  analogWrite(m2speed, 0);
}

void rotateRight(int duration, int pwm)
{
  digitalWrite(m1direction,HIGH); // forward
  digitalWrite(m2direction,LOW); // backward
  analogWrite(m1speed, pwm); // speed
  analogWrite(m2speed, pwm);
  delay(duration);
  analogWrite(m1speed, 0); // speed
  analogWrite(m2speed, 0);
}
```

❶ (at `digitalWrite(m1direction,HIGH); // forward` in goForward)

❷ (at `digitalWrite(m2direction,LOW); // backward` in goBackward)

❸ (at `digitalWrite(m1direction,HIGH); // forward` in rotateRight)

```
    void rotateLeft(int duration, int pwm)
    {
❹    digitalWrite(m1direction,LOW); // backward
     digitalWrite(m2direction,HIGH); // forward
     analogWrite(m1speed, pwm); // speed
     analogWrite(m2speed, pwm);
     delay(duration);
     analogWrite(m1speed, 0); // speed
     analogWrite(m2speed, 0);
    }

    void loop()
    {
      goForward(1000, 255);
      rotateLeft(1000, 255);
      goForward(1000, 255);
      rotateRight(1000, 255);
      goForward(1000, 255);
      goBackward(2000, 255);
      delay(2000);
    }
```

In the sketch, we set the direction of travel for each motor using

```
digitalWrite(m1direction,direction);
```

The value for *direction* is HIGH for forward or LOW for backward. Therefore, to make the tank move forward, we set both motors the same way, which has been done at ❶ and ❷. Next we set the speed of the motor using the following:

```
analogWrite(m1speed, pwm);
```

The value for *pwm* is the speed, between 0 and 255. To make the tank rotate left or right, the motors must be set in opposite directions, as shown at ❸ and ❹.

WARNING *When you're ready to upload the sketch, position the tank either by holding it off your work surface or by propping it up so that its treads aren't in contact with a surface; if you don't do this, then when the sketch upload completes, the tank will burst into life and leap off your desk after 5 seconds!*

Upload the sketch, remove the USB cable, and connect the battery cable to the Arduino power socket. Then place the tank on carpet or a clean surface, and let it drive about. Experiment with the movement functions in Project 40 to control your tank; this will help you become familiar with the time delays and how they relate to distance traveled.

Sensing Collisions

Now that our tank can move, we can start to add basic intelligence, such as collision sensors that can tell the tank when it has bumped into something or that can measure the distance between the tank and an object in its path so that it can avoid a crash. We'll use three methods of collision avoidance: microswitches, infrared, and ultrasonic.

Project #41: Detecting Tank Bot Collisions with a Microswitch

A *microswitch* can act like the simple push button we used in Chapter 4, but the microswitch component is physically larger and includes a large metal bar that serves as the actuator (see Figure 12-21).

Figure 12-21: Microswitch

When using the microswitch, you connect one wire to the bottom contact and the other to the contact labeled "NO" (normally open) to ensure that current flows only when the bar is pressed. We'll mount the microswitch on the front of our tank, and when the tank hits an object, the bar will be pressed, causing current to flow and making the tank reverse direction or take another action.

The Schematic

The microswitch hardware is wired like a single push button, as shown in Figure 12-22.

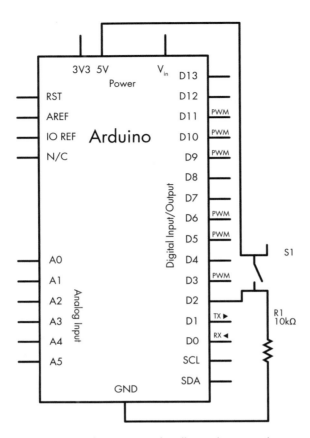

Figure 12-22: The microswitch collision detector schematic

The Sketch

We connect the microswitch to an interrupts port (digital pin 2). Although you might think we should have a function called by the interrupt to make the tank reverse for a few moments, that's not possible, because the delay() function doesn't operate inside functions called by interrupts. We have to think a little differently in this case.

Instead, the function goForward() will turn on the motors if two conditions are met for the variables crash and the Boolean move. If crash is true, the motors will reverse at a slower speed for 2 seconds to "back out" from a collision situation.

We can't use delay() functions because of the interrupt, so we measure the amount of time that the motors run reading millis() at the start and compare that against the current value of millis(). When the difference is greater than or equal to the required duration, move is set to false and the motors stop.

Enter and upload the following sketch:

```
// Project 41 - Detecting Tank Bot Collisions with a Microswitch

int m1speed=6; // digital pins for speed control
int m2speed=5;
int m1direction=7; // digital pins for direction control
int m2direction=4;
boolean crash=false;

void setup()
{
  pinMode(m1direction, OUTPUT);
  pinMode(m2direction, OUTPUT);
  attachInterrupt(0, backOut, RISING);
  delay(5000);
}
```

❶
```
void backOut()
{
  crash=true;
}
```

❷
```
void backUp()
{
    digitalWrite(m1direction,LOW); // reverse
    digitalWrite(m2direction,LOW); // reverse
    analogWrite(m1speed, 200); // speed
    analogWrite(m2speed, 200);
    delay(2000);
    analogWrite(m1speed, 0); // speed
    analogWrite(m2speed, 0);
}

void goForward(int duration, int pwm)
{
  long a,b;
  boolean move=true;
```
❸
```
  a=millis();
  do
  {
    if (crash==false)
    {
      digitalWrite(m1direction,HIGH); // forward
      digitalWrite(m2direction,HIGH); // forward
      analogWrite(m1speed, pwm); // speed
      analogWrite(m2speed, pwm);
    }
    if (crash==true)
    {
      backUp();
      crash=false;
    }
```

```
❹      b=millis()-a;
       if (b>=duration)
       {
         move=false;
       }
    } while (move!=false);
    // stop motors
    analogWrite(m1speed, 0);
    analogWrite(m2speed, 0);
}

void loop()
{
  goForward(5000, 255);
  delay(2000);
}
```

This sketch uses an advanced method of moving forward, in that two variables are used to monitor movement while the tank bot is in motion. The first is the Boolean variable crash. If the tank bot bumps into something and activates the microswitch, then an interrupt is called, which runs the function backOut() at ❶. It is here that the variable crash is changed from false to true. The second variable that is monitored is the Boolean variable move. In the function goForward(), we use millis() at ❸ to calculate constantly whether the tank bot has finished moving for the required period of time (set by the parameter duration).

At ❹, the function calculates whether the elapsed time is less than the required time, and if so, the variable move is set to true. Therefore, the tank bot is allowed to move forward only if it has not crashed and not run out of time. If a crash has been detected, the function backUp() at ❷ is called, at which point the tank will reverse slowly for 2 seconds and then resume as normal.

NOTE   *You can add the other movement functions from Project 40 to expand or modify this example.*

Infrared Distance Sensors

Our next method of collision avoidance uses an infrared (IR) distance sensor. This sensor bounces an infrared light signal off a surface in front of it and returns a voltage that is relative to the distance between the sensor and the surface. Infrared sensors are useful for collision detection because they are inexpensive, but they're not ideal for *exact* distance measuring. We'll use the Sharp GP2Y0A21YK0F analog sensor for our project, as shown in Figure 12-23.

Figure 12-23: The Sharp IR sensor

Wiring It Up

To wire the sensor, connect the red and black wires on the sensor to 5 V and GND, respectively, with the white wire connecting to an analog input pin on your Arduino. We'll use analogRead() to measure the voltage returned from the sensor. The graph in Figure 12-24 shows the relationship between the distance measured and the output voltage.

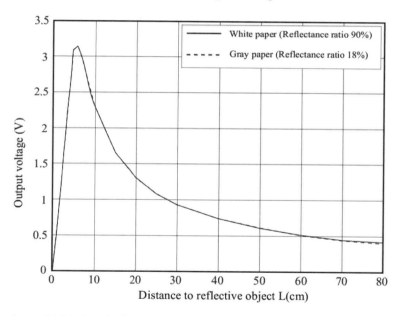

Figure 12-24: Graph of IR sensor distance versus output voltage

Testing the IR Distance Sensor

Because the relationship between distance and output is not easily represented with an equation, we'll categorize the readings into 5 cm stages. To demonstrate this, we'll use a simple example. Connect your infrared sensor's white lead to analog pin 0, the red lead to 5 V, and the black lead to GND, and then enter and upload the sketch shown in Listing 12-2.

```
// Listing 12-2

float sensor = 0;
int cm = 0;

void setup()
{
  Serial.begin(9600);
}

void loop()
{
  sensor = analogRead(0);
  if (sensor<=90)
  {
    Serial.println("Infinite distance!");
  } else if (sensor<100) // 80cm
  {
    cm = 80;
  } else if (sensor<110) // 70 cm
  {
    cm = 70;
  } else if (sensor<118) // 60cm
  {
    cm = 60;
  } else if (sensor<147) // 50cm
  {
    cm = 50;
  } else if (sensor<188) // 40 cm
  {
    cm = 40;
  } else if (sensor<230) // 30cm
  {
    cm = 30;
  } else if (sensor<302) // 25 cm
  {
    cm = 25;
  } else if (sensor<360) // 20cm
  {
    cm = 20;
  } else if (sensor<505) // 15cm
  {
    cm = 15;
  } else if (sensor<510) // 10 cm
  {
    cm = 10;
  } else if (sensor>=510) // too close!
  {
    Serial.println("Too close!");
  }
```

```
    Serial.print("Distance: ");
    Serial.print(cm);
    Serial.println(" cm");
    delay(250);
}
```

Listing 12-2: IR sensor demonstration sketch

The sketch reads the voltage from the IR sensor at ❶ and then uses a series of if statements at ❷ to choose which approximate distance is being returned. We determine the distance from the voltage returned by the sensor using two parameters. The first is the voltage-to-distance relationship as displayed in Figure 12-24. Then, using the knowledge (from Project 6) that analogRead() returns a value between 0 and 1,023 relative to a voltage between 0 and around 5 V, we can calculate the approximate distance returned by the sensor.

After uploading the sketch, open the Serial Monitor and experiment by moving your hand or a piece of paper at various distances from the sensor. The Serial Monitor should return the approximate distance, as shown in Figure 12-25.

Figure 12-25: Results of Listing 12-2

Project #42: Detecting Tank Bot Collisions with IR Distance Sensor

Now let's use the IR sensor with our tank bot instead of the microswitch. We'll use a slightly modified version of Project 41. Instead of using an interrupt, we'll create the function checkDistance() that changes the variable crash to true if the distance measured by the IR sensor is around 20 cm or less. We'll use this in the goForward() forward motion do... while loop.

Connect the IR sensor to your tank, and then enter and upload this sketch:

```
// Project 42 - Detecting Tank Bot Collisions with IR Distance Sensor

int m1speed=6;      // digital pins for speed control
int m2speed=5;
int m1direction=7; // digital pins for direction control
int m2direction=4;
boolean crash=false;

void setup()
{
  pinMode(m1direction, OUTPUT);
  pinMode(m2direction, OUTPUT);
  delay(5000);
}

void backUp()
{
    digitalWrite(m1direction,LOW); // reverse
    digitalWrite(m2direction,LOW); // reverse
    analogWrite(m1speed, 200);// speed
    analogWrite(m2speed, 200);
    delay(2000);
    analogWrite(m1speed, 0);// speed
    analogWrite(m2speed, 0);
}

void checkDistance()
{
❶   if (analogRead(0)>460)
    {
      crash=true;
    }
}

void goForward(int duration, int pwm)
{
  long a,b;
  boolean move=true;
  a=millis();
  do
  {
    checkDistance();
    if (crash==false)
    {
      digitalWrite(m1direction,HIGH); // forward
      digitalWrite(m2direction,HIGH); // forward
      analogWrite(m1speed, pwm); // speed
      analogWrite(m2speed, pwm);
    }
    if (crash==true)
    {
      backUp();
```

```
      crash=false;
    }
    b=millis()-a;
    if (b>=duration)
    {
      move=false;
    }
  } while (move!=false);
  // stop motors
  analogWrite(m1speed, 0);
  analogWrite(m2speed, 0);
}

void loop()
{
  goForward(5000, 255);
  delay(2000);
}
```

This sketch operates using the same methods used in Project 41, except this version constantly takes distance measurements at ❶ and sets the crash variable to true if the distance between the IR sensor and an object is less than about 20 cm.

After running the tank and using this sensor, you should see the benefits of using a noncontact collision sensor. It's simple to add more sensors to the same tank, such as sensors at the front and rear or at each corner. You should be able to add code to check each sensor in turn and make a decision based on the returned distance value.

Ultrasonic Distance Sensors

Our final method of collision avoidance is an *ultrasonic distance sensor*. This sensor bounces a sound wave of a frequency (that cannot be heard by the human ear) off a surface and measures the amount of time it takes for the sound to return to the sensor. We'll use the Parallax Ping))) ultrasonic distance sensor, shown in Figure 12-26, for this project, because it's inexpensive and accurate down to 1 cm.

Figure 12-26: The Ping))) ultrasonic distance sensor

An ultrasonic sensor's accuracy and range mean it can measure distances between 2 and 300 cm. However, because the sound wave needs to be reflected back to the sensor, the sensor must be angled less than 45 degrees away from the direction of travel.

Connecting the Ultrasonic Sensor

To connect the sensor, attach the 5 V and GND leads to their respective pins, and attach the SIG (short for *signal*) pin to any Arduino digital pin.

Using the Ultrasonic Sensor

The ultrasonic sensor takes measurements only when requested to do so. To take a measurement, we send a very short HIGH signal of 5 microseconds (mS) to the SIG pin. After a moment, the sensor should return a HIGH signal whose length is the period of time the ultrasonic sound takes to travel from and to the sensor; this value should be halved to determine the actual distance between the sensor and the object.

We need to use the same digital pin for output and input, and two new functions:

- `delayMicroseconds(mS)`   Pauses the Arduino sketch in microseconds (mS)
- `pulseDuration(pin, HIGH)`   Measures the length of a HIGH pulse on digital pin *pin* and returns the time in microseconds

After we have the duration of the incoming pulse, we convert it to centimeters by dividing it by 29.412 (because the speed of sound is 340 meters per second, or 34 cm per millisecond).

Testing the Ultrasonic Distance Sensor

To simplify using the sensor, we use the function getDistance() in Listing 12-3. Connect your ultrasonic sensor with the SIG pin to digital pin 3, and then enter and upload the following.

```
// Listing 12-3

int signal=3;

void setup()
{
  pinMode(signal, OUTPUT);
  Serial.begin(9600);
}

int getDistance()
```

```
// returns distance from Ping))) sensor in cm
{
  int distance;
  unsigned long pulseduration=0;

  // get the raw measurement data from Ping)))
  // set pin as output so we can send a pulse
❶  pinMode(signal, OUTPUT);

  // set output to LOW
  digitalWrite(signal, LOW);
  delayMicroseconds(5);

❷  // send the 5uS pulse out to activate Ping)))
  digitalWrite(signal, HIGH);
  delayMicroseconds(5);
  digitalWrite(signal, LOW);

❸  // change the digital pin to input to read the incoming pulse
  pinMode(signal, INPUT);

  // measure the length of the incoming pulse
  pulseduration=pulseIn(signal, HIGH);

❹  // divide the pulse length in half
  pulseduration=pulseduration/2;

❺  // convert to centimeters
  distance = int(pulseduration/29);
  return distance;
}

void loop()
{
  Serial.print(getDistance());
  Serial.println(" cm ");
  delay(500);
}
```

Listing 12-3: Ultrasonic sensor demonstration

The distance is returned by the function int getDistance(). By following ❶ through ❺, you can see how the pulse is sent to the sensor and then how the time of return is measured, which is used to calculate the distance.

After uploading the sketch, open the Serial Monitor and move an object toward and away from the sensor. The distance to the object should be returned in centimeters, as shown in Figure 12-27.

Figure 12-27: Results from Listing 12-3

Project #43: Detecting Tank Bot Collisions with an Ultrasonic Distance Sensor

Now that you understand how the sensor works, let's use it with our tank.

The Sketch

We can use the getDistance() function from Listing 12-3 to create a test for impending collision. In the following sketch, we check for distances of less than 10 cm, which will give the tank a reason to back up. Enter and upload the following sketch to see for yourself:

```
// Project 43 - Detecting Tank Bot Collisions with an Ultrasonic Distance
// Sensor

int m1speed=6; // digital pins for speed control
int m2speed=5;
int m1direction=7; // digital pins for direction control
int m2direction=4;
int signal=3;
boolean crash=false;

void setup()
{
  pinMode(m1direction, OUTPUT);
  pinMode(m2direction, OUTPUT);
  pinMode(signal, OUTPUT);
  delay(5000);
  Serial.begin(9600);
}
```

```
int getDistance()
// returns distance from Ping))) sensor in cm
{
  int distance;
  unsigned long pulseduration=0;

  // get the raw measurement data from Ping)))
  // set pin as output so we can send a pulse
  pinMode(signal, OUTPUT);

  // set output to LOW
  digitalWrite(signal, LOW);
  delayMicroseconds(5);

  // send the 5uS pulse out to activate Ping)))
  digitalWrite(signal, HIGH);
  delayMicroseconds(5);
  digitalWrite(signal, LOW);

  // change the digital pin to input to read the incoming pulse
  pinMode(signal, INPUT);

  // measure the length of the incoming pulse
  pulseduration=pulseIn(signal, HIGH);

  // divide the pulse length in half
  pulseduration=pulseduration/2;

  // convert to centimeters
  distance = int(pulseduration/29);
  return distance;
}

void backUp()
{
  digitalWrite(m1direction,LOW); // go back
  digitalWrite(m2direction,LOW);
  delay(2000);
  digitalWrite(m1direction,HIGH); // go left
  digitalWrite(m2direction,LOW);
  analogWrite(m1speed, 200); // speed
  analogWrite(m2speed, 200);
  delay(2000);
  analogWrite(m1speed, 0); // speed
  analogWrite(m2speed, 0);
}

void goForward(int duration, int pwm)
{
  long a,b;
  int dist=0;
  boolean move=true;
  a=millis();
  do
```

```
  {
    dist=getDistance();
    Serial.println(dist);
❶  if (dist<10) // if less than 10cm from object
    {
      crash=true;
    }
    if (crash==false)
    {
      digitalWrite(m1direction,HIGH); // forward
      digitalWrite(m2direction,HIGH); // forward
      analogWrite(m1speed, pwm); // speed
      analogWrite(m2speed, pwm);
    }
    if (crash==true)
    {
      backUp();
      crash=false;
    }
    b=millis()-a;
    if (b>=duration)
    {
      move=false;
    }
  } while (move!=false);
  // stop motors
  analogWrite(m1speed, 0);
  analogWrite(m2speed, 0);
}

void loop()
{
  goForward(1000, 255);
}
```

Once again, we constantly measure the distance at ❶ and then change the variable crash to true if the distance between the ultrasonic sensor and object is less than 10 cm. Watching the tank magically avoid colliding with things or having a battle of wits with a pet can be quite amazing.

Looking Ahead

In this chapter you learned how to introduce your Arduino-based projects to the world of movement. Using simple motors, or pairs of motors, with the motor shield, you can create projects that can move on their own and even avoid obstacles. We used three types of sensors to demonstrate a range of accuracies and sensor costs, so you can now make decisions based on your requirements and project budget.

By now, I hope you are experiencing and enjoying the ability to design and construct such things. But it doesn't stop here. In the next chapter, we move outdoors and harness the power of satellite navigation.

13

USING GPS
WITH YOUR ARDUINO

In this chapter, you will

- Learn how to connect a GPS shield
- Create a simple GPS coordinates display
- Show the actual position of GPS coordinates on a map
- Build an accurate clock
- Record the position of a moving object over time

You'll learn how to use an inexpensive GPS shield to determine location, create an accurate clock, and also make a logging device that records the position over time onto a microSD card, which can then be plotted over a map to display the movement history.

What Is GPS?

The *Global Positioning System (GPS)* is a satellite-based navigation system that sends data from satellites orbiting Earth to GPS receivers on the ground that can use that data to determine position and the current time anywhere on Earth. You are probably already familiar with GPS navigation devices used in cars or on your cell phone.

Although we can't create detailed map navigation systems with our Arduino, we can use a GPS module to determine position, time, and your approximate speed (if you're traveling). We'll do this by using the EM506 GPS receiver module shown in Figure 13-1.

Figure 13-1: EM506 GPS receiver

To use this receiver, you'll need the GPS shield kit from SparkFun (part number KIT-13199) shown in Figure 13-2. This part includes the receiver, a matching Arduino shield, and the connection cable.

Figure 13-2: Complete GPS shield bundle

You should also purchase a 1-foot-long cable (SparkFun part number GPS-09123), as shown in Figure 13-3, which will make placement of the receiver much simpler.

Figure 13-3: A long GPS-receiver-to-shield cable

Testing the GPS Shield

After you buy the GPS kit, it's a good idea to make sure that it's working and that you can receive GPS signals. GPS receivers require a line of sight to the sky, but their signals can pass through windows. It's usually best to perform this test outdoors, but it might work just fine through an unobstructed window or skylight. To test reception, you'll set up the shield and run a basic sketch that displays the raw received data.

To perform the test, first connect your GPS receiver via the cable to the shield, and then attach this assembly to your Arduino. Notice the small switch on the GPS shield, which is shown in Figure 13-4.

Figure 13-4: GPS shield data switch

When it's time to upload and run sketches on your GPS shield, move the switch to the DLINE position, and then change it to UART and be sure to turn on the power switch for the receiver.

Enter and upload the sketch in Listing 13-1.

```
// Listing 13-1

void setup()
{
❶  Serial.begin(4800);
}

void loop()
{
  byte a;
❷  if ( Serial.available() > 0 )
  {
    a = Serial.read();                // get the byte of data from the GPS
❸    Serial.write(a);
  }
}
```

Listing 13-1: Basic test sketch

This sketch listens to the serial port at ❷, and when a byte of data is received from the GPS module, it is sent to the Serial Monitor at ❸. (Notice that we start the Serial at 4,800 bps at ❶ to match the data speed of the GPS receiver.)

Once you've uploaded the sketch, move the data switch back to UART. Now check the LED on the GPS receiver that tells us its status. If the LED is lit, the GPS is trying to lock onto the satellite signals. After about 30 seconds the LED should start blinking, which indicates that it is receiving data from the satellite. After the LED starts blinking, open the Serial Monitor window in the IDE and set the data speed to 4,800 baud. You should see a constant stream of data similar to the output shown in Figure 13-5.

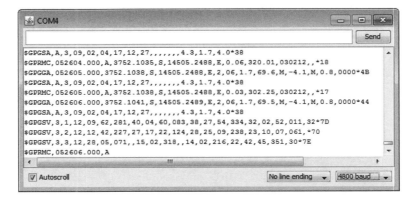

Figure 13-5: Raw data from GPS satellites

The data is sent from the GPS receiver to the Arduino one character at a time, and then it is sent to the Serial Monitor. But this isn't very useful as is, so we need to use a new library that extracts information from this raw data and converts it to a usable form. To do this, download and install the *TinyGPS* library from *http://www.arduiniana.org/libraries/tinygps/* using the method described in "Expanding Sketches with Libraries" on page 169.

Project #44: Creating a Simple GPS Receiver

Now let's create a simple GPS receiver. But first, because you'll usually use your GPS outdoors—and to make things a little easier—we'll add an LCD module to display the data, similar to the one shown in Figure 13-6.

NOTE   *Our examples are based on using the Freetronics LCD & Keypad shield. For more information on this shield, see* http://www.freetronics.com/collections/ display/products/lcd-keypad-shield/. *If you choose to use a different display module, be sure to substitute the correct values into the* LiquidCrystal *function in your sketches.*

Figure 13-6: The Freetronics LCD & Keypad shield

To display the current position coordinates received by the GPS on the LCD, we'll create a very basic portable GPS that could be powered by a 9 V battery and connector.

The Hardware

The required hardware is minimal:

- Arduino and USB cable
- LCD module or Freetronics LCD shield (mentioned earlier)
- One 9 V battery to DC socket cable
- One SparkFun GPS shield kit

The Sketch

Enter and upload the following sketch:

```
// Project 44 - Creating a Simple GPS Receiver
❶ #include <TinyGPS.h>
#include <LiquidCrystal.h>
LiquidCrystal lcd( 8, 9, 4, 5, 6, 7 );

// Create an instance of the TinyGPS object
TinyGPS gps;
```

```
❷ void getgps(TinyGPS &gps);

  void setup()
  {
    Serial.begin(4800);
    lcd.begin(16, 2);
  }

  void getgps(TinyGPS &gps)
  // The getgps function will display the required data on the LCD
  {
    float latitude, longitude;
    //decode and display position data
❸   gps.f_get_position(&latitude, &longitude);
    lcd.setCursor(0,0);
    lcd.print("Lat:");
    lcd.print(latitude,5);
    lcd.print("   ");
    lcd.setCursor(0,1);
    lcd.print("Long:");
    lcd.print(longitude,5);
    lcd.print("   ");
    delay(3000); // wait for 3 seconds
    lcd.clear();
  }

  void loop()
  {
    byte a;
    if ( Serial.available() > 0 )  // if there is data coming into the serial line
    {
      a = Serial.read();          // get the byte of data
      if(gps.encode(a))           // if there is valid GPS data...
      {
❹       getgps(gps);              // grab the data and display it on the LCD
      }
    }
  }
```

From ❶ to ❷, the sketch introduces the required libraries for the
LCD and GPS. In void loop at ❹, we send the characters received from the
GPS receiver to the function getgps() at ❸, which uses gps.f_get_position()
to insert the position values in the byte variables &latitude and &longitude
(which we display on the LCD).

Displaying the Position on the LCD

After the sketch has been uploaded and the GPS starts receiving data, your
current position in decimal latitude and longitude should be displayed on
your LCD, as shown in Figure 13-7.

Figure 13-7: Latitude and longitude display from Project 44

But where on Earth is this? We can determine exactly where it is by using Google Maps (*http://maps.google.com/*). On the website, enter the latitude and longitude, separated by a comma and a space, into the search field, and Google Maps will return the location. For example, using the coordinates returned in Figure 13-7 produces a map like the one shown in Figure 13-8.

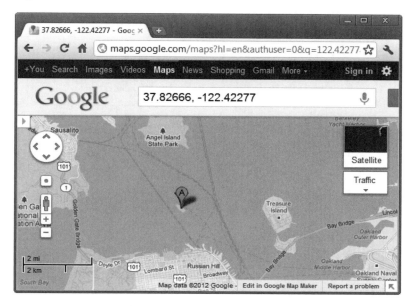

Figure 13-8: Location of position displayed in Figure 13-7

Project #45: Creating an Accurate GPS-based Clock

There is more to a GPS than finding a location; the system also transmits time data that can be used to make a very accurate clock.

The Hardware

For this project, we'll use the same hardware from Project 44.

The Sketch

Enter the following sketch to build a GPS clock:

```
// Project 45 - Creating an Accurate GPS-based Clock
#include <TinyGPS.h>
#include <LiquidCrystal.h>
LiquidCrystal lcd( 8, 9, 4, 5, 6, 7 );

// Create an instance of the TinyGPS object
TinyGPS gps;

void getgps(TinyGPS &gps);

void setup()
{
  Serial.begin(4800);
  lcd.begin(16, 2);
}

void getgps(TinyGPS &gps)
{
  int year,a,t;
  byte month, day, hour, minute, second, hundredths;
❶  gps.crack_datetime(&year,&month,&day,&hour,&minute,&second,&hundredths);

❷  hour=hour+10;                        // correct for your time zone
  if (hour>23)
  {
    hour=hour-24;
  }
  lcd.setCursor(0,0);                   // print the date and time
❸  lcd.print("Current time: ");
  lcd.setCursor(4,1);
  if (hour<10)
  {
    lcd.print("0");
  }
  lcd.print(hour, DEC);
  lcd.print(":");
  if (minute<10)
  {
    lcd.print("0");
  }
  lcd.print(minute, DEC);
  lcd.print(":");
  if (second<10)
  {
    lcd.print("0");
  }
  lcd.print(second, DEC);
}
```

```
void loop()
{
  byte a;
  if ( Serial.available() > 0 )   // if there is data coming into the serial line
  {
    a = Serial.read();            // get the byte of data
    if(gps.encode(a))             // if there is valid GPS data...
    {
      getgps(gps);                // then grab the data and display it on the LCD
    }
  }
}
```

This example works in a similar way to Project 44, except that instead of extracting the position data, the sketch extracts the time (always at Greenwich Mean Time, more commonly known as UTC) at ❶. At ❷, you can either add or subtract a number of hours to bring the clock into line with your current time zone. The time should then be formatted clearly and displayed on the LCD at ❸. Figure 13-9 shows an example of the clock.

Figure 13-9: Project 45 at work

Project #46: Recording the Position of a Moving Object over Time

Now that you know how to receive GPS coordinates and convert them into normal variables, we can use this information with the microSD card shield from Chapter 8 to build a GPS logger. Our logger will record our position over time by logging the GPS data over time. The addition of the microSD card shield will allow you to record the movement of a car, truck, boat, or any other moving object that allows GPS signal reception; later, you can review the information on a computer.

The Hardware

The required hardware is the same as that used for the previous examples, except that you need to replace the LCD shield with the microSD shield from Chapter 8, and you'll use external power. In our example, we'll record the time, position information, and estimated speed of travel.

The Sketch

After assembling your hardware, enter and upload the following sketch:

```
// Project 46 - Recording the Position of a Moving Object over Time
#include <SD.h>
#include <TinyGPS.h>

// Create an instance of the TinyGPS object
TinyGPS gps;

void getgps(TinyGPS &gps);

void setup()
{
  pinMode(10, OUTPUT);
  Serial.begin(9600);
  // check that the microSD card exists and can be used
  if (!SD.begin(8)) {
    Serial.println("Card failed, or not present");
    // stop the sketch
    return;
  }
  Serial.println("microSD card is ready");
}

void getgps(TinyGPS &gps)
{
  float latitude, longitude;
  int year;
  byte month, day, hour, minute, second, hundredths;

  //decode and display position data
  gps.f_get_position(&latitude, &longitude);
  File dataFile = SD.open("DATA.TXT", FILE_WRITE);
  // if the file is ready, write to it
  if (dataFile)
  {
    dataFile.print("Lat: ");
    dataFile.print(latitude,5);
    dataFile.print(" ");
    dataFile.print("Long: ");
    dataFile.print(longitude,5);
    dataFile.print(" ");
    // decode and display time data
    gps.crack_datetime(&year,&month,&day,&hour,&minute,&second,&hundredths);
    // correct for your time zone as in Project 45
    hour=hour+11;
    if (hour>23)
    {
      hour=hour-24;
    }
```

❶ `if (dataFile)`

❷ `dataFile.print("Lat: ");`

```
    if (hour<10)
    {
      dataFile.print("0");
    }
    dataFile.print(hour, DEC);
    dataFile.print(":");
    if (minute<10)
    {
      dataFile.print("0");
    }
    dataFile.print(minute, DEC);
    dataFile.print(":");
    if (second<10)
    {
      dataFile.print("0");
    }
    dataFile.print(second, DEC);
    dataFile.print(" ");
    dataFile.print(gps.f_speed_kmph());
❸    dataFile.println("km/h");
    dataFile.close();
❹    delay(30000);  // record a measurement every 30 seconds
  }
}

void loop()
{
  byte a;
  if ( Serial.available() > 0 )  // if there is data coming into the serial line
  {
    a = Serial.read();           // get the byte of data
    if(gps.encode(a))            // if there is valid GPS data...
    {
❺      getgps(gps);             // then grab the data and display it on the LCD
    }
  }
}
```

This sketch uses the same code used in Projects 44 and 45 in void loop() to receive data from the GPS receiver and pass it on to other functions. At ❺, the text from the GPS receiver is passed into the *TinyGPS* library to decode the data into useful variables. At ❶, the microSD card is checked to determine whether data can be written to it, and from ❷ to ❸, the relevant GPS data is written to the text file on the microSD card. Because the file is closed after every write, you can remove the power source from the Arduino without warning the sketch, and you should do so before inserting or removing the microSD card. Finally, you can set the interval between recording data at ❹ by changing the value in the delay() function.

Displaying Locations on a Map

After operating your GPS logger, the resulting text file should look similar to Figure 13-10.

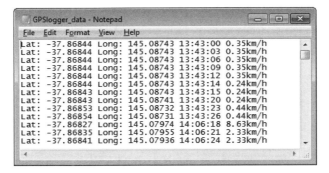

Figure 13-10: Results from Project 46

Once you have this data, you can enter it into Google Maps manually and review the path taken by the GPS logger, point by point. But a more interesting method is to display the entire route taken on one map. To do this, open the text file as a spreadsheet, separate the position data, and add a header row, as shown in Figure 13-11. Then save it as a *.csv* file.

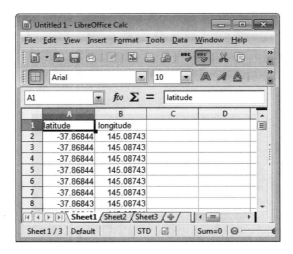

Figure 13-11: Captured position data

Now visit the GPS Visualizer website (*http://www.gpsvisualizer.com/*). In the Get Started Now box, click **Choose File** and select your data file. Choose **Google Maps** as the output format, and then click **Go!**. The movement of your GPS logger should be shown on a map similar to the one in Figure 13-12, which you can then adjust and explore.

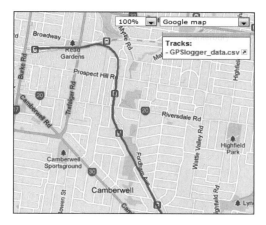

Figure 13-12: Mapped GPS logger data

Looking Ahead

As you can see, something that you might have thought too complex, such as working with GPS receivers, can be made pretty simply with your Arduino. Continuing with that theme, in the next chapter you'll learn how to create your own wireless data links and control things via remote control.

14

WIRELESS DATA

In this chapter you'll learn how to send and receive instructions and data using various types of wireless transmission hardware. Specifically, you'll learn how to

- Send digital output signals using low-cost wireless modules
- Create a simple and inexpensive wireless remote control system
- Use XBee wireless data receivers and transceivers
- Create a remote control temperature sensor

Using Low-cost Wireless Modules

It's easy to send text information in one direction using a wireless link between two Arduino-controlled systems that have inexpensive radio frequency (RF) data modules, such as the transmitter module shown in Figure 14-1. These modules are usually sold in pairs and are known as *RF Link* modules or kits. Good examples are part WLS107B4B from Seeed Studio and part WRL-10534 from SparkFun. We'll use the most common module types that run on the 433 MHz radio frequency in our projects.

The connections shown at the bottom of the module in Figure 14-1 are, from left to right, 5 V, GND, data in, and an external antenna. The antenna can be a single length of wire, or it can be omitted entirely for short transmission distances. (Each brand of module can vary slightly, so check the connections on your particular device before moving forward.)

Figure 14-2 shows the receiver module, which is slightly larger than the transmitter module.

Figure 14-1: Transmitter RF Link module

Figure 14-2: Receiver RF Link module

The connections on the receiver are straightforward: the V+ and V– pins connect to 5 V and GND, respectively, and DATA connects to the Arduino pin allocated to receive the data.

Before you can use these modules, you also need to download and install the latest version of the *VirtualWire* library from *http://www.airspayce .com/mikem/arduino/VirtualWire/*. After you've installed the library, you'll be ready to move on to the next section.

NOTE   *The RF Link modules are inexpensive and easy to use, but they have no error-checking capability to ensure that data being sent is received correctly. Therefore, I recommend that you use them only for simple tasks such as this basic remote control project. If your project calls for more accurate and reliable data transmission, use something like the XBee modules instead, which are discussed later in this chapter.*

Project #47: Creating a Wireless Remote Control

We'll remotely control two digital outputs: You'll press buttons connected to one Arduino board to control matching digital output pins on another Arduino located some distance away. This project will show you how to use the RF Link modules and how to determine what sort of distance you can achieve remote control with the modules before you commit to using the modules for more complex tasks. (In open air, the distance you can achieve is generally about 100 meters, but the distance will be less when you are indoors or when the modules are between obstacles.)

The Hardware for the Transmitter Circuit

The following hardware is required for the transmitter circuit:

- Arduino and USB cable
- One 9 V battery and DC socket cable (as used in Chapter 12)
- One 433 MHz RF Link transmitter module (such as SparkFun part number WRL-10534)
- Two 10 kΩ resistors (R1 and R2)
- Two 100 nF capacitors (C1 and C2)
- Two push buttons
- One breadboard

The Transmitter Schematic

The transmitter circuit consists of two push buttons with debounce circuitry connected to digital pins 2 and 3, and the transmitter module wired as described earlier, as shown in Figure 14-3.

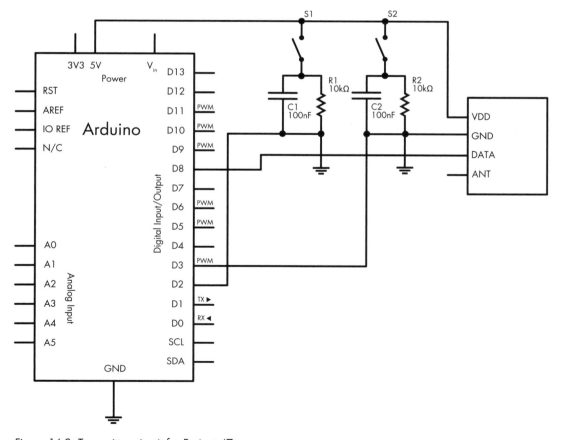

Figure 14-3: Transmitter circuit for Project 47

The Hardware for the Receiver Circuit

The following hardware is required for the receiver circuit:

- Arduino and USB cable
- One 9 V battery and DC socket cable (as used in Chapter 12)
- One 433 MHz RF Link receiver module (such as SparkFun part number WRL-10532)
- One breadboard
- Two LEDs of your choice
- Two 560 Ω resistors (R1 and R2)

The Receiver Schematic

The receiver circuit consists of two LEDs on digital pins 6 and 7, and the data pin from the RF Link receiver module connected to digital pin 8, as shown in Figure 14-4.

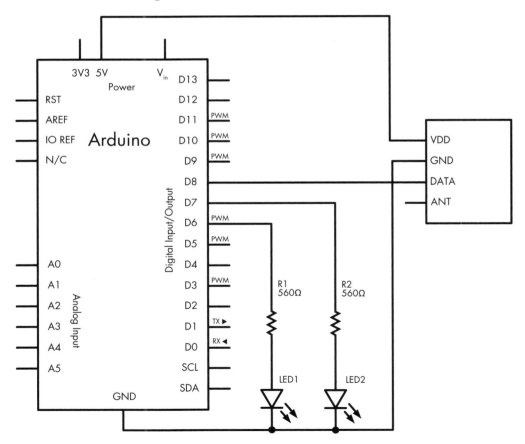

Figure 14-4: Receiver circuit for Project 47

You can substitute the breadboard, LEDs, resistors, and receiver module with a Freetronics 433 MHz receiver shield, shown in Figure 14-5.

Figure 14-5: Freetronics 433 MHz receiver shield

The Transmitter Sketch

Now let's examine the sketch for the transmitter. Enter and upload the following sketch to the Arduino with the transmitter circuit:

```
// Project 47 - Creating a Wireless Remote Control, Transmitter Sketch

❶ #include <VirtualWire.h>
  uint8_t buf[VW_MAX_MESSAGE_LEN];
  uint8_t buflen = VW_MAX_MESSAGE_LEN;

❷ const char *on2 = "a";
  const char *off2 = "b";
  const char *on3 = "c";
  const char *off3 = "d";

  void setup()
  {
❸     vw_set_ptt_inverted(true);      // Required for RF Link modules
      vw_setup(300);                  // set data speed
❹     vw_set_tx_pin(8);
      pinMode(2, INPUT);
      pinMode(3, INPUT);
  }

  void loop()
  {
❺   if (digitalRead(2)==HIGH)
    {
        vw_send((uint8_t *)on2, strlen(on2));  // send the data out to the world
        vw_wait_tx();                          // wait a moment
        delay(200);
    }
    if (digitalRead(2)==LOW)
    {
❻       vw_send((uint8_t *)off2, strlen(off2));
        vw_wait_tx();
        delay(200);
    }
    if (digitalRead(3)==HIGH)
    {
        vw_send((uint8_t *)on3, strlen(on3));
        vw_wait_tx();
        delay(200);
    }
```

```
    if (digitalRead(3)==LOW)
    {
        vw_send((uint8_t *)off3, strlen(off3));
        vw_wait_tx();
        delay(200);
    }
}
```

To use the RF Link modules, we use the virtual wire functions at ❶ and ❸ in the sketch. At ❹, we set digital pin 8, which is used to connect the Arduino to the data pin of the transmitter module and to control the speed of the data transmission. (You can use any other digital pins if necessary, except 0 and 1, which would interfere with the serial line.)

The transmitter sketch reads the status of the two buttons connected to digital pins 2 and 3 and sends a single text character to the RF Link module that matches the state of the buttons. For example, when the button on digital pin 2 is HIGH, the Arduino sends the character *a*, and when the button is LOW, it sends the character *b*. The four states are declared starting at ❷.

The transmission of the text character is handled using one of the four sections' if statements, starting at ❺—for example, the contents of the if-then statement at ❻. The variable transmitted is used twice—for example, on2, as shown here:

```
vw_send((uint8_t *)on2, strlen(on2));
```

The function vw_send sends the contents of the variable on2, but it needs to know the length of the variable in characters, so we use strlen() to accomplish this.

The Receiver Sketch

Now let's add the receiver sketch. Enter and upload the following sketch to the receiver circuit's Arduino:

```
// Project 47 - Creating a Wireless Remote Control, Receiver Sketch
#include <VirtualWire.h>

uint8_t buf[VW_MAX_MESSAGE_LEN];
uint8_t buflen = VW_MAX_MESSAGE_LEN;

void setup()
{
❶  vw_set_ptt_inverted(true);     // Required for RF link modules
    vw_setup(300);
❷  vw_set_rx_pin(8);
    vw_rx_start();
    pinMode(6, OUTPUT);
    pinMode(7, OUTPUT);
}
```

```
      void loop()
      {
❸     if (vw_get_message(buf, &buflen))
      {
❹       switch(buf[0])
        {
          case 'a':
          digitalWrite(6, HIGH);
          break;
          case 'b':
          digitalWrite(6, LOW);
          break;
          case 'c':
          digitalWrite(7, HIGH);
          break;
          case 'd':
          digitalWrite(7, LOW);
          break;
        }
      }
      }
```

As with the transmitter circuit, we use the virtual wire functions at ❶ to set up the RF Link receiver module, set the data speed, and set the Arduino digital pin to which the link's data output pin is connected at ❷.

When the sketch is running, the characters sent from the transmitter circuit are received by the RF Link module and sent to the Arduino. The function vw_get_message() at ❸ takes the characters received by the Arduino, which are interpreted by the switch-case function at ❹. For example, pressing button S1 on the transmitter circuit will send the character *a*. This character is received by the transmitter, which sets digital pin 6 to HIGH, turning on the LED.

You can use this simple pair of demonstration circuits to create more complex controls for Arduino systems by sending codes as basic characters to be interpreted by a receiver circuit.

Using XBee Wireless Data Modules for Greater Range and Faster Speed

When you need a wireless data link with greater range and a faster data speed than what the basic wireless modules used earlier can provide, XBee data modules may be the right choice. These modules transmit and receive serial data directly to one or more XBee modules and to and from a computer. Several models of XBee modules are available, but we'll use the Series 1 line of XBee transceivers, shown in Figure 14-6.

Figure 14-6: Typical XBee transceiver

Connecting the transceiver to an Arduino is simple with an XBee shield, shown in Figure 14-7.

Figure 14-7: XBee connected to an Arduino via a typical XBee shield

To communicate with a computer, you can use the XBee Explorer board, such as the board shown in Figure 14-8 (SparkFun part WRL-08687).

Figure 14-8: XBee connected via USB
with an Explorer board

XBees don't require their own library; they operate as a simple serial *data bridge* that sends and receives data via the Arduino's serial line.

Project #48: Transmitting Data with an XBee

This project will demonstrate simple data transmission by sending data from an Arduino to a computer that has an XBee and Explorer board. The project involves two basics steps: First, we'll connect an XBee to an XBee shield. Then we'll connect the shield to the Arduino.

Notice the tiny switch on the Arduino shield, as shown in Figure 14-9.

This switch is identical to the switch on the GPS shield used in Chapter 13. This data switch controls whether data from the Arduino USB or the XBee is sent to the microcontroller. When you're uploading a sketch, set the switch to DLINE; you'll change it to UART when you're running a sketch.

Figure 14-9: XBee shield data switch

The Sketch

Create and upload the following sketch:

```
// Project 48 - Transmitting Data with an XBee

void setup()
{
  Serial.begin(9600);
}

void loop()
{
  Serial.println("Hello, world");
  delay(1000);
}
```

Although this code simply sends the text *Hello, world* to the serial line, it nicely demonstrates the simplicity of using XBee modules to send data: You simply write to the serial line, and the XBee takes care of the rest. Remove the USB cable from the Arduino board, and power the board using an external supply such as the 9 V battery and cable from Chapter 12. (Don't forget to set the switch back to UART.)

Setting Up the Computer to Receive Data

Now let's set up the computer to receive data. Using an Explorer board (Figure 14-8), connect another XBee to your computer. Then download and install the Terminal program, available for Windows from *https://sites .google.com/site/terminalbpp/*. When you open Terminal, you should see a screen like the one shown in Figure 14-10.

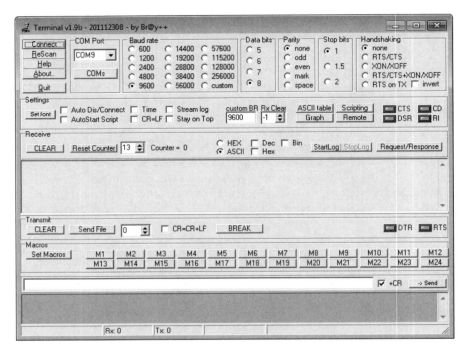

Figure 14-10: Terminal emulator software

At the top of the screen, set the Baud Rate to **9600**, Data Bits to **8**, Parity to **none**, Stop Bits to **1**, and Handshaking to **none**. Then click the **ReScan** button in the upper-left corner to force the software to select the correct USB port. Finally, click **Connect**, and after a moment you should see data being sent from the Arduino in the Terminal window, as shown in Figure 14-11.

Figure 14-11: Reception of data from the remote XBee

You can attach multiple XBee transmitters to one computer, with each sending its own data for monitoring purposes. For example, from the comfort of your desk, you could monitor various sensors, such as temperature sensors, placed throughout an area. You could even control multiple XBee-connected Arduinos using a similar method, as you'll see in the next project.

Project #49: Building a Remote Control Thermometer

In this project, you'll build a remote control thermometer that returns the temperature upon request from your computer. In the process, you'll learn the basics of remote-sensor data retrieval and create a base for use with other projects that involve remote sensing.

The Hardware

The following hardware is required:

- Arduino and USB cable
- One TMP36 temperature sensor
- Two XBee (Series 1) modules
- One XBee Arduino shield
- One XBee Explorer board and USB cable
- Various connecting wires
- A 9 V battery and DC socket cable (as used in Chapter 12)
- One breadboard

The Layout

Insert one XBee into the shield; then connect the shield to the Arduino board. Use the small solderless breadboard to hold the TMP36 and add wires to 5 V, GND, and A0 on the XBee shield (as with Project 8), as shown in Figure 14-12.

Figure 14-12: Remote temperature-sensing Arduino setup

The USB Explorer board and the other XBee will be connected to the PC after you upload the following sketch to your Arduino board.

The Sketch

Enter and upload the following sketch:

```
// Project 49 - Building a Remote Control Thermometer

char a;
float voltage=0;
float sensor=0;
float celsius=0;
float fahrenheit=0;
float photocell=0;

void setup()
{
  Serial.begin(9600);
}

void sendC()
{
  sensor=analogRead(0);
  voltage=((sensor*5000)/1024);
  voltage=voltage-500;
  celsius=voltage/10;
  fahrenheit=((celsius*1.8)+32);
  Serial.print("Temperature: ");
  Serial.print(celsius,2);
  Serial.println(" degrees C");
}

void sendF()
{
  sensor=analogRead(0);
  voltage=((sensor*5000)/1024);
  voltage=voltage-500;
  celsius=voltage/10;
  fahrenheit=((celsius*1.8)+32);
  Serial.print("Temperature: ");
  Serial.print(fahrenheit,2);
  Serial.println(" degrees F");
}

void getCommand()
{
  Serial.flush();
  while (Serial.available() == 0)
  {
    // do nothing until data arrives from XBee
  }
  while (Serial.available() > 0)
  {
    a = Serial.read(); // read the number in the serial buffer
  }
}
```

❶

```
  void loop()
  {
    getCommand();// listen for command from PC
❷  switch (a)
    {
❸   case 'c':
      // send temperature in Celsius
      sendC();
      break;
❹   case 'f':
      // send temperature in Fahrenheit
      sendF();
      break;
    }
  }
```

First, the sketch listens to the serial line using the function getCommand()
at ❶. It will loop around until it receives a character, at which point it stores
that character in variable a. Once the character has been received, it is
interpreted using the switch-case function at ❷. If the letter *c* is received, the
function sendC() at ❸ calculates the temperature in Celsius, writes it to the
serial line, and then sends the temperate to the host PC via the XBees. If
the letter *f* is received, sendF() at ❹ calculates the temperature in Fahrenheit
and sends it back to the PC.

Operation

Run the Terminal program, as you did earlier in this chapter, and then
send out a *c* or an *f*. The temperature should be returned in about 1 second
and should look similar to the results shown in Figure 14-13.

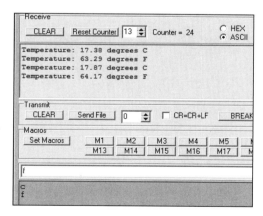

Figure 14-13: Results from Project 49

Looking Ahead

This chapter shows how simple it is to control multiple Arduino systems remotely. For example, you could control digital outputs by sending characters from the PC terminal or another Arduino, or have certain characters recognized by individual Arduino boards. With the knowledge you've gained so far, many more creative options are available to you.

But there's still much more to investigate in terms of wireless data transmission, so keep reading and working along as you learn to use simple television remote controls with the Arduino in the next chapter.

15

INFRARED REMOTE CONTROL

In this chapter you will

- Create and test a simple infrared receiver
- Remotely control Arduino digital output pins
- Add a remote control system to the motorized tank we created in Chapter 12

As you'll see, with the addition of an inexpensive receiver module, your Arduino can receive the signals from an infrared remote and act upon them.

What Is Infrared?

Although most people use infrared remote controls in a variety of daily actions, they don't know how they work. Infrared (IR) signals are actually beams of light that operate at a wavelength that cannot be seen by the naked eye. So when you look at the little LED poking out of a remote control and press a button, you won't see the LED light up.

IR remote controls contain one or more special infrared light–generating LEDs that are used to transmit the IR signals. For example, when you press a button on the remote, the LED turns on and off repeatedly in a pattern that is unique for each button pressed on the remote. This signal is received by a special IR receiver and converted to pulses of electrical current that are then converted to data in the receiver's electronics. If you are curious about these patterns, you can actually view them by looking at the IR LED on a remote through the viewfinder of a phone camera or digital camera.

Setting Up for Infrared

Before moving forward, we need to install the infrared Arduino library, so visit *https://github.com/shirriff/Arduino-IRremote/* to download the required files and install them, as explained in "Expanding Sketches with Libraries" on page 169.

The IR Receiver

The next step is to set up the IR receiver and test that it is working. You can choose either an independent IR receiver (shown in Figure 15-1) or a prewired module (shown in Figure 15-2), which is the easier route to take.

Figure 15-1: IR receiver

Figure 15-2: Pre-wired IR receiver module

The independent IR receiver shown in Figure 15-1 is a Vishay TSOP4138, available from retailers such as Newark (part number 59K0287) or element14 (part number 1040743). The bottom leg of the receiver (as shown in the figure) connects to an Arduino digital pin, the center leg to GND, and the top leg to 5V.

Figure 15-2 shows a prewired IR module. Prewired receiver modules are available from DFRobot and other retailers. The benefit of using these modules is that they include connection wires and are labeled for easy reference.

Regardless of your choice of module, in all of the following examples, you'll connect the D (or data line) to Arduino digital pin 11, VCC to 5V, and GND to GND.

The Remote Control

Finally, you will need a remote control. We'll use a Sony TV remote like the one shown in Figure 15-3. If you don't have access to a Sony remote, any inexpensive universal remote control can be used after you reset it to Sony codes. See the instructions included with your remote control to do this.

Figure 15-3: Typical Sony remote control

A Test Sketch

Now let's make sure that everything works. After connecting your IR receiver to the Arduino, enter and upload the sketch in Listing 15-1.

```
// Listing 15-1

   int receiverpin = 11;      // pin 1 of IR receiver to Arduino digital pin 11
❶  #include <IRremote.h>      // use the library
❷  IRrecv irrecv(receiverpin); // create instance of irrecv
❸  decode_results results;

   void setup()
   {
     Serial.begin(9600);
     irrecv.enableIRIn();      // start the IR receiver
   }

   void loop()
   {
❹    if (irrecv.decode(&results)) // have we received an IR signal?
     {
❺      Serial.print(results.value, HEX); // display IR code on the Serial Monitor
       Serial.print(" ");
       irrecv.resume(); // receive the next value
     }
   }
```

Listing 15-1: IR receiver test

This sketch is relatively simple, because most of the work is done in the background by the IR library. At ❹, we check to see if a signal has been received from the remote control. If so, it is displayed on the Serial Monitor in hexadecimal at ❺. The lines at ❶, ❷, and ❸ activate the IR library and create an instance of the infrared library function to refer to in the rest of the sketch.

Testing the Setup

Once you've uploaded the sketch, open the Serial Monitor, aim the remote at the receiver, and start pressing buttons. You should see codes displayed on the Serial Monitor after each button press. For example, Figure 15-4 shows the results of pressing 1, 2, and 3, once each.

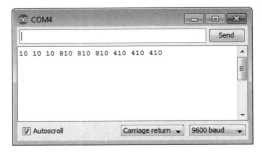

Figure 15-4: Results of pressing buttons after running the code in Listing 15-1

Table 15-1 lists the codes from a basic Sony remote control that we'll use in upcoming sketches. However, when running Listing 15-1, notice how each code number repeats three times. This is an idiosyncrasy of Sony IR systems, which send the code three times for each button press. You can ignore these repeats with some clever coding, but for now let's skip to remote controlling with the next project.

Table 15-1: Example Sony IR codes

Button	Code	Button	Code
Power	A90	7	610
Mute	290	8	E10
1	10	9	110
2	810	0	910
3	410	Volume up	490
4	C10	Volume down	C90
5	210	Channel up	90
6	A10	Channel down	890

Project #50: Creating an IR Remote Control Arduino

This project will demonstrate how to control digital output pins using IR remote control. This project will allow you to control digital pins 2 through 7 with the matching numerical buttons 2 through 7 on a Sony remote control. When you press a button on the remote control, the matching digital output pin will change state to HIGH for 1 second and then return to LOW. You'll be able to use this project as a base or guide to add remote control capabilities to your other projects.

The Hardware

The hardware is the same as that required for the IR test at the beginning of this chapter, with the addition of the custom LED shield you made for Project 28. (You could always connect LEDs, motors, or other forms of output instead of this LED shield.)

The Sketch

Enter and upload the following sketch:

```
// Project 50 - Creating an IR Remote Control Arduino

int receiverpin = 11;        // pin 1 of IR receiver to Arduino digital pin 11
#include <IRremote.h>
IRrecv irrecv(receiverpin); // create instance of irrecv
decode_results results;

void setup()
{
  irrecv.enableIRIn(); // start the receiver
  for (int z = 2 ; z < 8 ; z++) // set up digital pins
  {
    pinMode(z, OUTPUT);
  }
}

❶ void translateIR()
  // takes action based on IR code received
  // uses Sony IR codes
  {
    switch(results.value)
    {
❷    case 0x810:  pinOn(2); break; // 2
      case 0x410:  pinOn(3); break; // 3
      case 0xC10:  pinOn(4); break; // 4
      case 0x210:  pinOn(5); break; // 5
      case 0xA10:  pinOn(6); break; // 6
      case 0x610:  pinOn(7); break; // 7
    }
  }
```

```
❸ void pinOn(int pin) // turns on digital pin "pin" for 1 second
  {
    digitalWrite(pin, HIGH);
    delay(1000);
    digitalWrite(pin, LOW);
  }

  void loop()
  {
❹  if (irrecv.decode(&results)) // have we received an IR signal?
    {
      translateIR();
❺    for (int z = 0 ; z < 2 ; z++) // ignore the 2nd and 3rd repeated codes
      {
        irrecv.resume(); // receive the next value
      }
    }
  }
```

This sketch has three major parts. First, it waits for a signal from the remote at ❹. When a signal is received, the signal is tested in the function translateIR() at ❶ to determine which button was pressed and what action to take.

Notice at ❷ how we compare the hexadecimal codes returned by the IR library. These are the codes returned by the test conducted in Listing 15-1. When the codes for buttons 2 through 7 are received, the function pinOn() at ❸ is called, which turns on the matching digital pin for 1 second.

As mentioned, Sony remotes send the code three times for each button press, so we use a small loop at ❺ to ignore the second and third codes. Finally, note the addition of 0x in front of the hexadecimal numbers used in the case statements at ❷.

NOTE   *Hexadecimal numbers are base 16 and use the digits 0 through 9 and then A through F, before moving on to the next column. For example, decimal 10 in hexadecimal is A, decimal 15 in hexadecimal is F, decimal 16 is 10 hexadecimal, and so on. When using a hexadecimal number in a sketch, preface it with 0x.*

Expanding the Sketch

You can expand the options or controls available for controlling your motorized tank by testing more buttons. To do so, use Listing 15-1 to determine which button creates which code, and then add each new code to the switch...case statement.

Project #51: Creating an IR Remote Control Tank

To show you how to integrate an IR remote control into an existing project, we'll add IR to the tank described for Project 40. In this project, instead of presetting the tank's direction and distances, the sketch will show you how to control these actions with a simple Sony TV remote.

The Hardware

The required hardware is the same as that required for the tank you built for Project 40, with the addition of the IR receiver module in the method described earlier in this chapter. In the following sketch, the tank will respond to the buttons that you press on the remote control as follows: press 2 for forward, 8 for backward, 4 for rotate left, and 6 for rotate right.

The Sketch

After reassembling your tank and adding the IR receiver, enter and upload the following sketch:

```
// Project 51 - Creating an IR Remote Control Tank

int receiverpin = 11;        // pin 1 of IR receiver to Arduino digital pin 11
#include <IRremote.h>

IRrecv irrecv(receiverpin); // create instance of 'irrecv'
decode_results results;

int m1speed    = 6; // digital pins for speed control
int m2speed    = 5;
int m1direction = 7; // digital pins for direction control
int m2direction = 4;

void setup()
{
  pinMode(m1direction, OUTPUT);
  pinMode(m2direction, OUTPUT);
  irrecv.enableIRIn(); // start IR receiver
}

void goForward(int duration, int pwm)
{
  digitalWrite(m1direction, HIGH); // forward
  digitalWrite(m2direction, HIGH); // forward
  analogWrite(m1speed, pwm);       // at selected speed
  analogWrite(m2speed, pwm);
  delay(duration);                 // and duration
  analogWrite(m1speed, 0);         // then stop
  analogWrite(m2speed, 0);
}
```

```
void goBackward(int duration, int pwm)
{
  digitalWrite(m1direction, LOW);   // backward
  digitalWrite(m2direction, LOW);   // backward
  analogWrite(m1speed, pwm);        // at selected speed
  analogWrite(m2speed, pwm);
  delay(duration);
  analogWrite(m1speed, 0);          // then stop
  analogWrite(m2speed, 0);
}

void rotateRight(int duration, int pwm)
{
  digitalWrite(m1direction, HIGH); // forward
  digitalWrite(m2direction, LOW);  // backward
  analogWrite(m1speed, pwm);       // at selected speed
  analogWrite(m2speed, pwm);
  delay(duration);                 // and duration
  analogWrite(m1speed, 0);         // then stop
  analogWrite(m2speed, 0);
}

void rotateLeft(int duration, int pwm)
{
  digitalWrite(m1direction, LOW);  // backward
  digitalWrite(m2direction, HIGH); // forward
  analogWrite(m1speed, pwm);       // at selected speed
  analogWrite(m2speed, pwm);
  delay(duration);                 // and duration
  analogWrite(m1speed, 0);         // then stop
  analogWrite(m2speed, 0);
}

// translateIR takes action based on IR code received, uses Sony IR codes
void translateIR()
{
  switch(results.value)
  {
    case 0x810:  goForward(250, 255);    break; // 2
    case 0xC10:  rotateLeft(250, 255);   break; // 4
    case 0xA10:  rotateRight(250, 255);  break; // 6
    case 0xE10:  goBackward(250, 255);   break; // 8
  }
}
```

```
void loop()
{
  if (irrecv.decode(&results)) // have we received an IR signal?
  {
    translateIR();
    for (int z = 0 ; z < 2 ; z++) // ignore the repeated codes
    {
      irrecv.resume(); // receive the next value
    }
  }
}
```

This sketch should look somewhat familiar to you. Basically, instead of turning on digital pins, it calls the motor control functions that were used in the tank from Chapter 12.

Looking Ahead

Having worked through the projects in this chapter, you should understand how to send commands to your Arduino via an infrared remote control device. Combined with your existing knowledge (and forthcoming projects), you now can replace physical forms of input such as buttons with a remote control.

But the fun doesn't stop here. In the next chapter, we'll use Arduino to harness something that, to the untrained eye, is fascinating and futuristic: radio-frequency identification systems.

16

READING RFID TAGS

In this chapter you will

- Learn how to implement RFID readers with your Arduino
- Understand how to save variables in the Arduino EEPROM
- Design the framework for an Arduino-based RFID access system

Radio-frequency identification, or *RFID*, is a wireless system that uses electromagnetic fields to transfer data from one object to another, without the two objects touching. You can build an Arduino that reads common RFID tags and cards to create access systems and to control digital outputs. You may have used an RFID card before, such as an access card that you use to unlock a door or a public transport card that you hold in front of a reader on the bus. Figure 16-1 shows some examples of RFID tags and cards.

Figure 16-1: Example RFID devices

Inside RFID Devices

Inside an RFID device is a tiny integrated circuit with memory that can be accessed by a specialized reader. Most tags don't have a battery inside; instead, they're powered by energy from an electromagnetic field produced by the RFID reader. This field is transmitted by a fine coil of wire, which also acts as the antenna for the transmission of data between the card and the reader. Figure 16-2 shows the antenna coil of the RFID reader that we'll use in this chapter.

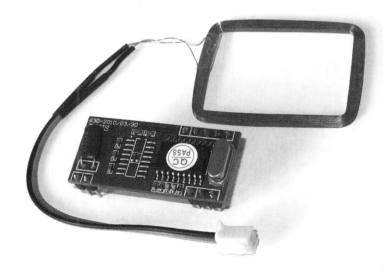

Figure 16-2: Our RFID reader

The card reader we'll use in this chapter is available from Seeed Studio at *http://www.seeedstudio.com/*, part number RFR101A1M. It's cheap and easy to use, and it operates at 125 kHz; be sure to purchase RFID tags that match that frequency, such as Seeed Studio part number RFR103B2B.

Testing the Hardware

In this section you'll connect the RFID reader to the Arduino and then test that it's working with a simple sketch that reads RFID cards and sends the data to the Serial Monitor.

The Schematic

Figure 16-3 shows the RFID module connections.

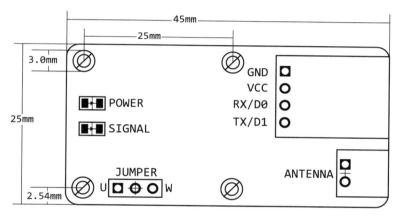

Figure 16-3: RFID module connections

Testing the Schematic

Once the RFID reader is connected to the Arduino, you can test it by placing the black jumper across the left and center pins of the jumper section. Then, to make the connections between the RFID reader and the Arduino, follow these steps, using female-to-male jumper wires:

1. Connect the included coil plug to the antenna socket.
2. Connect the reader GND to the Arduino GND.
3. Connect VCC to Arduino 5V.
4. Connect RX to Arduino pin D0.
5. Connect TX to Arduino pin D1.

NOTE *When uploading sketches to the RFID-connected Arduino, be sure to remove the wire between the RFID reader RX and Arduino pin D0. Then reconnect it after the sketch has been uploaded successfully. This is because the D0 pin is also used by the Arduino board to communicate and receive sketches.*

The Test Sketch

Enter and upload Listing 16-1.

```
// Listing 16-1
int data1 = 0;

void setup()
{
  Serial.begin(9600);
}

void loop()
{
  if (Serial.available() > 0) {
    data1 = Serial.read();
    // display incoming number
    Serial.print(" ");
    Serial.print(data1, DEC);
  }
}
```

Listing 16-1: RFID test sketch

Check Serial Monitor

Open the Serial Monitor window and wave an RFID tag over the coil. The results should look similar to Figure 16-4.

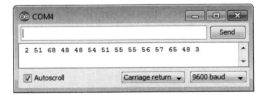

Figure 16-4: Example output from Listing 16-1

Notice that 14 numbers are displayed in the Serial Monitor window. This is the RFID tag's unique ID number, which we'll use in future sketches to identify the tag being read. Now record the numbers that result from each of your tags, because you'll need them for the next few projects.

Project #52: Creating a Simple RFID Control System

Now let's put the RFID system to use. In this project you'll learn how to trigger an Arduino event when the correct RFID tag is read. This sketch stores two RFID tag numbers; when one of the cards is read by the reader, it will display *Accepted* in the Serial Monitor. If the wrong card is presented, then the Serial Monitor will display *Rejected*. We'll use this as a base to add RFID controls to existing projects.

The Sketch

Enter and upload the following sketch. However, at ❶ and ❷, replace the *x*s in each array with the numbers for each of the RFID tags you generated earlier in the chapter. (We discussed arrays in Chapter 6.)

```
// Project 52 - Creating a Simple RFID Control System

int data1 = 0;
int ok = -1;

// use Listing 16-1 to find your tag numbers
❶ int tag1[14] = {x, x, x, x, x, x, x, x, x, x, x, x, x, x};
❷ int tag2[14] = {x, x, x, x, x, x, x, x, x, x, x, x, x, x};

int newtag[14] = {0,0,0,0,0,0,0,0,0,0,0,0,0,0}; // used for read comparisons

void setup()
{
  Serial.flush(); // need to flush serial buffer,
                  // otherwise first read may not be correct
  Serial.begin(9600);
}

❸ boolean comparetag(int aa[14], int bb[14])
  {
    boolean ff = false;
    int fg = 0;
    for (int cc = 0 ; cc < 14 ; cc++)
    {
      if (aa[cc] == bb[cc])
      {
        fg++;
      }
    }
    if (fg == 14)
    {
      ff = true;
    }
    return ff;
  }
```

```
❹ void checkmytags()  // compares each tag against the tag just read
  {
    ok = 0; // this variable helps decision-making,
            // if it is 1 we have a match, zero is a read but no match,
            // -1 is no read attempt made
    if (comparetag(newtag, tag1) == true)
    {
❺    ok++;
    }
    if (comparetag(newtag, tag2) == true)
    {
❻    ok++;
    }
  }

  void loop()
  {
    ok = -1;
    if (Serial.available() > 0) // if a read has been attempted
    {
      // read the incoming number on serial RX
      delay(100);  // needed to allow time for the data
                   // to come in from the serial buffer.
      for (int z = 0 ; z < 14 ; z++) // read the rest of the tag
❼     {
        data1 = Serial.read();
        newtag[z] = data1;
      }
      Serial.flush(); // stops multiple reads
      // now to match tags up
❽     checkmytags();
    }
❾   //now do something based on tag type
    if (ok > 0) // if we had a match
    {
      Serial.println("Accepted");
      ok = -1;
    }
    else if (ok == 0) // if we didn't have a match
    {
      Serial.println("Rejected");
      ok = -1;
    }
  }
```

How It Works

When a tag is presented to the RFID reader, it sends the tag numbers
through the serial port. We capture all 14 of these numbers and place
them in the array newtag[] at ❼. Next, the tag is compared against the

two stored tag numbers at ❶ and ❷ using the function checkmytags() at ❹ and ❽, with the actual comparisons of the tag arrays passed to the function comparetag() at ❸.

The comparetag() function accepts the two number arrays as parameters and returns (in Boolean) whether the arrays are identical (true) or different (false). If a match is made, the variable OK is set to 1 at ❺ and ❻. Finally, at ❾, we have the actions to take once the tag read succeeds.

After uploading the sketch and reconnecting the wire from Arduino D0 to the RFID reader RX (see Figure 16-3), open the Serial Monitor window and present some tags to the reader. The results should be similar to Figure 16-5.

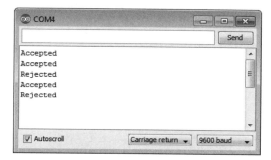

Figure 16-5: Results of Project 52

Storing Data in the Arduino's Built-in EEPROM

When you define and use a variable in your Arduino sketches, the stored data lasts only until the Arduino is reset or the power is turned off. But what if you want to keep the values for future use, such as in the user-changeable secret code for the numeric keypad in Chapter 9? That's where the *EEPROM (electrically erasable read-only memory)* comes in. The EEPROM stores variables in memory inside an ATmega328 microcontroller, and the variables aren't lost when the power is turned off.

The EEPROM in the Arduino can store 1,024-byte variables in positions numbered from 0 to 1,023. Recall that a byte can store an integer with a value between 0 and 255, and you begin to see why it's perfect for storing RFID tag numbers. To use the EEPROM in our sketches, we first call the EEPROM library (included with the Arduino IDE) using the following:

```
#include <EEPROM.h>
```

Then, to write a value to the EEPROM, we simply use this:

```
EEPROM.write(a, b);
```

Here, a is the position in which the value (which falls between 0 and 1,023) will be stored, and b is the variable holding the byte of data we want to store in the EEPROM position number a.

To retrieve data from the EEPROM, use this function:

```
value = EEPROM.read(position);
```

This takes the data stored in EEPROM position number position and stores it in the variable value.

NOTE   *The EEPROM has a finite life, and it can eventually wear out! According to the manufacturer, Atmel, it can sustain 100,000 write/erase cycles for each position. Reads are unlimited.*

Reading and Writing to the EEPROM

Here's an example of how to read and write to the EEPROM. Enter and upload Listing 16-2.

```
// Listing 16-2
#include <EEPROM.h>
int zz;

void setup()
{
  Serial.begin(9600);
  randomSeed(analogRead(0));
}

void loop()
{
  Serial.println("Writing random numbers...");
  for (int i = 0; i < 1024; i++)
  {
    zz = random(255);
❶    EEPROM.write(i, zz);
  }
  Serial.println();
  for (int a = 0; a < 1024; a++)
  {
❷    zz = EEPROM.read(a);
    Serial.print("EEPROM position: ");
    Serial.print(a);
    Serial.print(" contains ");
❸    Serial.println(zz);
    delay(25);
  }
}
```

Listing 16-2: EEPROM demonstration sketch

In the loop at ❶, a random number between 0 and 255 is stored in each EEPROM position. The stored values are retrieved in the second loop at ❷, to be displayed in the Serial Monitor at ❸.

Once the sketch has been uploaded, open the Serial Monitor and you should see something like Figure 16-6.

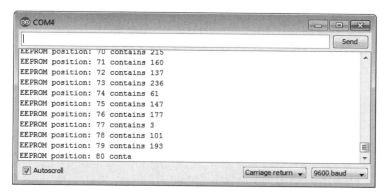

Figure 16-6: Example output from Listing 16-2

Now you're ready to create a project using the EEPROM.

Project #53: Creating an RFID Control with "Last Action" Memory

Although Project 52 showed how to use an RFID to control something, such as a light or electric door lock, we had to assume that nothing would be remembered if the system were reset or the power went out. For example, if a light was on and the power went out, then the light would be off once the power returned. However, once the power comes back on, you want the Arduino to remember what was happening before the power went out and return to that state. Let's solve that problem now.

In this project, the last action will be stored in the EEPROM (for example, "locked" or "unlocked"). When the sketch restarts after a power failure or an Arduino reset, the system will revert to the previous state stored in the EEPROM.

The Sketch

Enter and upload the following sketch. Again, replace the xs in each array at ❶ and ❷ with the numbers for each of the two RFID tags as you did for Project 52.

```
// Project 53 - Creating an RFID Control with "Last Action" Memory

#include <EEPROM.h>
```

```
  int data1 = 0;
  int ok = -1;
  int lockStatus = 0;

  // use Listing 16-1 to find your tag numbers
❶ int tag1[14] = {x, x, x, x, x, x, x, x, x, x, x, x, x, x};
❷ int tag2[14] = {x, x, x, x, x, x, x, x, x, x, x, x, x, x};

  int newtag[14] = {0,0,0,0,0,0,0,0,0,0,0,0,0,0}; // used for read comparisons

  void setup()
  {
    Serial.flush();
    Serial.begin(9600);
    pinMode(13, OUTPUT);
❸   checkLock();
  }

  // comparetag compares two arrays and returns true if identical
  // this is good for comparing tags
  boolean comparetag(int aa[14], int bb[14])
  {
    boolean ff = false;
    int fg = 0;
    for (int cc = 0; cc < 14; cc++)
    {
      if (aa[cc] == bb[cc])
      {
        fg++;
      }
    }
    if (fg == 14)
    {
      ff = true;
    }
    return ff;
  }

  void checkmytags()
  // compares each tag against the tag just read
  {
    ok = 0;
    if (comparetag(newtag, tag1) == true)
    {
      ok++;
    }
    if (comparetag(newtag, tag2) == true)
    {
      ok++;
    }
  }
```

❹ ```
void checkLock()
{
  Serial.print("System Status after restart ");
  lockStatus = EEPROM.read(0);
  if (lockStatus == 1)
  {
    Serial.println("- locked");
    digitalWrite(13, HIGH);
  }
  if (lockStatus == 0)
  {
    Serial.println("- unlocked");
    digitalWrite(13, LOW);
  }
  if ((lockStatus != 1) && (lockStatus != 0))
  {
    Serial.println("EEPROM fault - Replace Arduino hardware");
  }
}

void loop()
{
  ok = -1;
  if (Serial.available() > 0) // if a read has been attempted
  {
    // read the incoming number on serial RX
    delay(100);
    for (int z = 0; z < 14; z++) // read the rest of the tag
    {
      data1 = Serial.read();
      newtag[z] = data1;
    }
    Serial.flush(); // prevents multiple reads
    // now to match tags up
    checkmytags();
  }
```
❺ ```
  if (ok > 0) // if we had a match
  {
    lockStatus = EEPROM.read(0);
    if (lockStatus == 1) // if locked, unlock it
```
❻ ```
    {
      Serial.println("Status - unlocked");
      digitalWrite(13, LOW);
      EEPROM.write(0, 0);
    }
    if (lockStatus == 0)
```
❼ ```
    {
      Serial.println("Status - locked");
      digitalWrite(13, HIGH);
      EEPROM.write(0, 1);
    }
```

```
        if ((lockStatus != 1) && (lockStatus != 0))
❽      {
          Serial.println("EEPROM fault - Replace Arduino hardware");
        }
      }
      else if (ok == 0) // if we didn't have a match
      {
        Serial.println("Incorrect tag");
        ok = -1;
      }
      delay(500);
    }
```

How It Works

This sketch is a modification of Project 52. We use the onboard LED to simulate the status of something that we want to turn on or off every time an acceptable RFID tag is read. After a tag has been read and matched, the status of the lock is changed at ❺. We store the status of the lock in the first position of the EEPROM. The status is represented by a number: 0 is unlocked and 1 is locked. This status will change (from locked to unlocked and back to locked) after every successful tag read at ❻ or ❼.

We've also introduced a fail-safe in case the EEPROM has worn out. If the value returned from reading the EEPROM is not 0 or 1, we should be notified at ❽. Furthermore, the status is checked when the sketch restarts after a reset using the function checkLock() at ❶, ❷, ❸, and ❹, which reads the EEPROM value, determines the last status, and then sets the lock to that status (locked or unlocked).

Looking Ahead

Once again we have used an Arduino board to re-create simply what could be a very complex project. You now have a base to add RFID control to your projects to create professional-quality access systems and to control digital outputs with the swipe of an RFID card. We'll demonstrate this again when we revisit RFID in Chapter 18.

17

DATA BUSES

In this chapter you will

- Learn about the I^2C bus
- Understand how to use an EEPROM (electrically erasable read-only memory) and a port expander on the I^2C bus
- Learn about the SPI bus
- Learn how to use a digital rheostat on the SPI bus

An Arduino communicates with other devices via a *data bus*, a system of connections that allow two or more devices to exchange data in an orderly manner. A data bus can provide a connection between the Arduino and various sensors, I/O expansion devices, and other components.

The two major buses of most importance to the Arduino are the *Serial Peripheral Interface (SPI)* bus and the *Inter-Integrated Circuit bus (I^2C)*. Many useful sensors and external devices communicate using these buses.

The I²C Bus

The I²C bus, also known as the *Two Wire Interface (TWI)* bus, is a simple and easy device used for data communication. Data is transferred between devices and the Arduino through two wires, known as *SDA* and *SCL* (the data line and clock line, respectively). In the case of the Arduino Uno, the SDA pin is A4 and the SCL pin is A5, as shown in Figure 17-1. Some newer R3 boards also have dedicated I²C pins at the upper-left corner for convenient access.

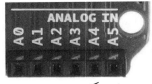

Figure 17-1: The I²C bus connectors on the Arduino Uno

On the I²C bus, the Arduino is considered the *master device*, and each IC out on the bus is a *slave*. Each slave has its own address, a hexadecimal number that allows the Arduino to address and communicate with each device. Each device usually has a range of I²C bus addresses to choose from, which is detailed in the manufacturer's data sheet. The particular addresses available are determined by wiring the IC pins a certain way.

NOTE   *Because the Arduino runs on 5 V, your I²C device must also operate on 5 V or be able to tolerate it. Always confirm this by contacting the seller or manufacturer before use.*

To use the I²C bus, you'll need to use the `Wire` library (included with the Arduino IDE):

```
#include <Wire.h>
```

Next, in `void setup()`, activate the bus with this:

```
Wire.begin();
```

Data is transmitted along the bus 1 byte at a time. To send a byte of data from the Arduino to a device on the bus, three functions are required:

1. The first function initiates communication with the following line of code (where *address* is the slave's bus address in hexadecimal—for example 0x50):

```
Wire.beginTransmission(address);
```

2. The second function sends 1 byte of data from the Arduino to the device addressed by the previous function (where `data` is a variable containing 1 byte of data; you can send more than 1 byte, but you'll need to use one `Wire.write()` for each byte):

```
Wire.write(data);
```

3. Finally, once you have finished sending data to a particular device, use this to end the transmission:

```
Wire.endTransmission();
```

To request that data from an I²C device be sent to the Arduino, start with Wire.beginTransmission(*address*), followed by the following (where *x* is the number of bytes of data to request):

```
Wire.requestFrom(address,x);
```

Next, use the following function to store the incoming byte into a variable:

```
incoming = Wire.read(); // incoming is the variable receiving the byte of data
```

Now finalize the transaction with Wire.endTransmission(), and we'll put these functions to use in the next project.

Project #54: Using an External EEPROM

In Chapter 16 we used the Arduino's internal EEPROM to prevent the erasure of variable data caused by a board reset or power failure. The Arduino's internal EEPROM stores only 1,024 bytes of data. To store more data, you can use external EEPROMs, as you'll see in this project.

For our external EEPROM, we'll use the Microchip Technology 24LC512 EEPROM, which can store 64KB (65,536 bytes) of data (Figure 17-2). It's available from retailers such as Digi-Key (part number 24LC512-I/P-ND) and element14 (part number 1660008).

Figure 17-2: Microchip Technology's 24LC512 EEPROM

The Hardware

Here's what you'll need to create this project:

- One Microchip Technology 24LC512 EEPROM
- One breadboard
- Two 4.7 kΩ resistors
- One 100 nF ceramic capacitor
- Various connecting wires
- Arduino and USB cable

The Schematic

For the circuit, connect each 4.7 kΩ resistor between 5V and SCL and between 5V and SDA, as shown in Figure 17-3.

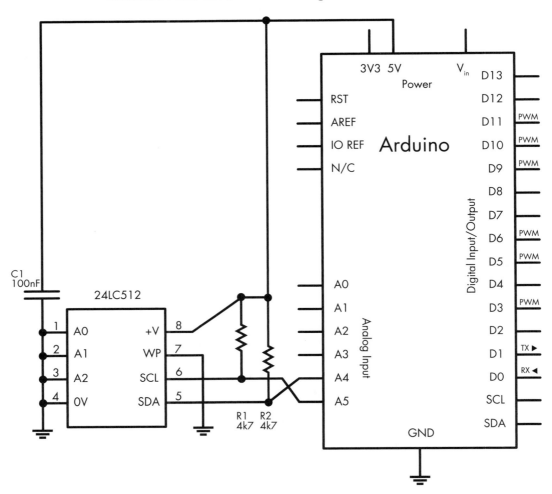

Figure 17-3: Schematic for Project 54

The bus address for the 24LC512 EEPROM IC is partially determined by the way it is wired into the circuit. The last 3 bits of the bus address are determined by the status of pins A2, A1, and A0. When these pins are connected to GND, their values are 0; when they are connected to 5V, their values are 1.

The first 4 bits are preset as 1010. Therefore, in our circuit, the bus address is represented as 1010000 in binary, which is 0x50 in hexadecimal. This means that we can use 0x50 as the bus address in the sketch.

The Sketch

Although our external EEPROM can store up to 64KB of data, our sketch is intended to demonstrate just a bit of its use, so we'll store and retrieve bytes only in the EEPROM's first 20 memory positions.

Enter and upload the following sketch:

```
// Project 54 - Using an External EEPROM
#include <Wire.h>
#define chip1 0x50

unsigned int pointer;
byte d=0;

void setup()
{
  Serial.begin(9600);
  Wire.begin();
}

void writeData(int device, unsigned int address, byte data)
// writes a byte of data 'data' to the EEPROM at I2C address 'device'
// in memory location 'address'
{
  Wire.beginTransmission(device);
  Wire.write((byte)(address >> 8));   // left part of pointer address
  Wire.write((byte)(address & 0xFF)); // and the right
  Wire.write(data);
  Wire.endTransmission();
  delay(10);
}

byte readData(int device, unsigned int address)
// reads a byte of data from memory location 'address'
// in chip at I2C address 'device'
{
  byte result;  // returned value
  Wire.beginTransmission(device);
  Wire.write((byte)(address >> 8));   // left part of pointer address
  Wire.write((byte)(address & 0xFF)); // and the right
  Wire.endTransmission();
  Wire.requestFrom(device,1);
  result = Wire.read();
  return result; // and return it as a result of the function readData
}

void loop()
{
  Serial.println("Writing data...");
  for (int a=0; a<20; a++)
  {
    writeData(chip1,a,a);
  }
```

```
Serial.println("Reading data...");
for (int a=0; a<20; a++)
{
  Serial.print("EEPROM position ");
  Serial.print(a);
  Serial.print(" holds ");
  d=readData(chip1,a);
  Serial.println(d, DEC);
}
}
```

Let's walk through the sketch. At ❶, we activate the library and define the I²C bus address for the EEPROM as chip1. At ❷, we start the Serial Monitor and then the I²C bus. The two custom functions writeData() and readData() are included to save you time and give you some reusable code for future work with this EEPROM IC. We'll use them to write and read data, respectively, from the EEPROM.

The function writeData() at ❸ initiates transmission with the EEPROM, sends the address of where to store the byte of data in the EEPROM using the next two Wire.write() functions, sends a byte of data to be written, and then ends transmission.

The function readData() at ❹ operates the I²C bus in the same manner as writeData(), but instead of sending a byte of data to the EEPROM, it uses Wire.requestFrom() to read the data at ❺. Finally, the byte of data sent from the EEPROM is received into the variable result and becomes the return value for the function.

The Result

In void loop() the sketch loops 20 times and writes a value to the EEPROM. Then it loops again, retrieving the values and displaying them in the Serial Monitor, as shown in Figure 17-4.

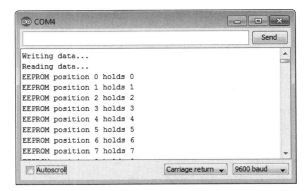

Figure 17-4: Results of Project 54

Project #55: Using a Port Expander IC

A *port expander* is another useful IC that is controlled via I²C. It's designed to offer more digital outputs. In this project, we'll use the Microchip Technology MCP23017 16-bit port expander IC (Figure 17-5), which has 16 digital outputs to add to your Arduino. It is available from retailers such as Newark (part number 31K2959) or element14 (part number 1332088).

Figure 17-5: Microchip Technology's MCP23017 port expander IC

In this project, we'll connect the MCP23017 to an Arduino and demonstrate how to control the 16 port expander outputs with the Arduino. Each of the port expander's outputs can be treated like a regular Arduino digital output.

The Hardware

Here's what you'll need to create this project:

- Arduino and USB cable
- One breadboard
- Various connecting wires
- One Microchip Technology MCP20317 port expander IC
- Two 4.7 kΩ resistors
- (Optional) An equal number of 560 Ω resistors and LEDs

The Schematic

Figure 17-6 shows the basic schematic for an MCP23017. As with the EEPROM from Project 54, we can set the I²C bus address by using a specific wiring order. With the MCP23017, we connected pins 15 through 17 to GND to set the address to 0x20.

When you're working with the MCP23017, it helps to have the pin-out diagram from the IC's data sheet, as shown in Figure 17-7. Note that the 16 outputs are divided into two banks: GPA7 through GPA0 on the right side and GPB0 through GPB7 on the left. We'll connect LEDs via 560 Ω resistors from some or all of the outputs to demonstrate when the outputs are being activated.

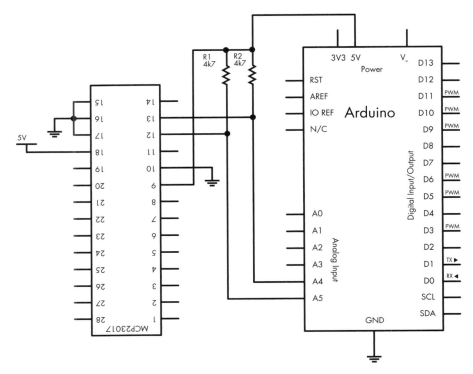

Figure 17-6: Basic schematic for Project 55

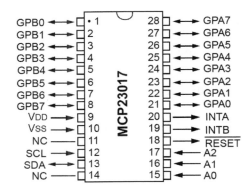

Figure 17-7: Pinout diagram for MCP23017

The Sketch

Enter and upload the following sketch:

```
// Project 55 - Using a Port Expander IC

#include "Wire.h"
#define mcp23017 0x20
```

```
void setup()
{
  Wire.begin(); // activate I2C bus
❶  // setup MCP23017
  // set I/O pins to outputs
  Wire.beginTransmission(mcp23017);
  Wire.write(0x00); // IODIRA register
  Wire.write(0x00); // set all of bank A to outputs
  Wire.write(0x00); // set all of bank B to outputs
❷  Wire.endTransmission();
}

void loop()
{
  Wire.beginTransmission(mcp23017);
  Wire.write(0x12);
❸  Wire.write(255);      // bank A
❹  Wire.write(255);      // bank B
  Wire.endTransmission();
  delay(1000);

  Wire.beginTransmission(mcp23017);
  Wire.write(0x12);
  Wire.write(0);      // bank A
  Wire.write(0);      // bank B
  Wire.endTransmission();
  delay(1000);
}
```

To use the MCP23017, we need the lines listed in void setup() from ❶ through ❷. To turn on and off the outputs on each bank, we send 1 byte representing each bank in order; that is, we send a value for bank GPA0 through GPA7 and then a value for GPB0 through GPB7.

When setting individual pins, you can think of each bank as a binary number (as explained in "A Quick Course in Binary" on page 116). Thus, to turn on pins 7 through 4, you would send the number 11110000 in binary (240 in decimal), inserted into the Wire.write() function shown at ❸ for bank GPA0 through GPA7 or ❹ for bank GPB0 through GPB7.

Hundreds of devices use the I^2C bus for communication. Now that you know the basics of how to use these buses, you can use any of them with an Arduino board.

The SPI Bus

The SPI bus differs from the I^2C bus in that it can be used to send data to and receive data from a device simultaneously and at different speeds, depending on the microcontroller used. Communication, however, is also *master-slave*: The Arduino acts as the *master* and determines which device (the *slave*) it will communicate with at one time.

Pin Connections

Each SPI device uses four pins to communicate with a master: *MOSI* (Master-Out, Slave-In), *MISO* (Master-In, Slave-Out), *SCK* (clock), and *SS* or *CS* (Slave Select or Chip Select). These SPI pins are connected to the Arduino as shown in Figure 17-8.

A typical single Arduino-to-SPI device connection is shown in Figure 17-9. Arduino pins D11 through D13 are reserved for SPI, but the SS pin can use any other digital pin (often D10 is used because it's next to the SPI pins).

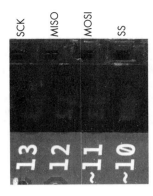

Figure 17-8: SPI pins on an Arduino Uno

Figure 17-9: Typical Arduino-to-SPI device connection

NOTE   *As with I²C devices, your SPI device must either operate on 5 V or tolerate it since the Arduino runs on 5 V. Be sure to check this out with the seller or manufacturer before use.*

Implementing the SPI

Now let's examine how to implement the SPI bus in a sketch. Before doing this, however, we'll run through the functions used. First is the SPI library (included with the Arduino IDE software):

```
#include "SPI.h"
```

Next, you need to choose a pin to be used for SS and set it as a digital output in void setup. Because we're using only one SPI device in our example, we'll use D10 and set it up HIGH first, because most SPI devices have an "active low" SS pin:

```
pinMode(10, OUTPUT);
digitalWrite(10, HIGH);
```

Here is the function to activate the SPI bus:

```
SPI.begin();
```

Finally, we need to tell the sketch which way to send and receive data. Some SPI devices require that their data be sent with the Most Significant Bit (MSB) first, and some want the MSB last. (Again, see "A Quick Course in Binary" on page 116 for more on MSB.) Therefore, in void setup we use the following function after SPI.begin:

```
SPI.setBitOrder(order);
```

Here, *order* is either MSBFIRST or MSBLAST.

Sending Data to an SPI Device

To send data to an SPI device, we first set the SS pin to LOW, which tells the SPI device that the master (the Arduino) wants to communicate with it. Next, we send bytes of data to the device with the following line, as often as necessary—that is, you use this once for each byte you are sending:

```
SPI.transfer(byte);
```

After you've finished communicating with the device, set the SS pin to HIGH to tell the device that the Arduino has finished communicating with it.

Each SPI device requires a separate SS pin. For example, if you had two SPI devices, the second SPI device's SS pin could be D9 and connected to the Arduino as shown in Figure 17-10.

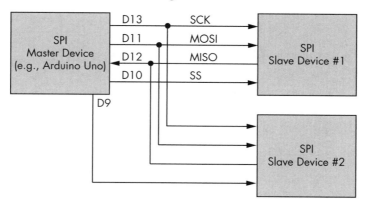

Figure 17-10: Two SPI devices connected to one Arduino

When communicating with the second slave device, you would use the D9 (instead of the D10) SS pin before and after communication.

Project 56 demonstrates using the SPI bus with a digital rheostat.

Project #56: Using a Digital Rheostat

In simple terms, a *rheostat* device is similar to the potentiometers we examined in Chapter 4, except the rheostat has two pins: one for the wiper and one for the return current. In this project, you'll use a digital rheostat to set the resistance in the sketch instead of physically turning a potentiometer knob or shaft yourself. Rheostats are often the basis of volume controls in audio equipment that use buttons rather than dials. The tolerance of a rheostat is much larger than that of a normal fixed-value resistor—in some cases, around 20 percent larger.

For Project 56, we will use the Microchip Technology MCP4162 shown in Figure 17-11. The MCP4162 is available in various resistance values; this example uses the 10 kΩ version. It is available from retailers such as Newark (part number 77M2766) and element14 (part number 1840698). The resistance can be adjusted in 255 steps; each step has a resistance of around 40 Ω. To select a particular step, we send 2 bytes of data to a command byte (which is 0) and the value byte (which is between 0 and 255). The MCP4162 uses nonvolatile memory, so once the power is disconnected and later connected, the last value selected is still in effect.

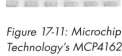

Figure 17-11: Microchip Technology's MCP4162 digital rheostat

We'll control the brightness of an LED using the rheostat.

The Hardware

Here's what you'll need to create this project:

- Arduino and USB cable
- One breadboard
- Various connecting wires
- One Microchip Technology MCP4162 digital rheostat
- One 560 Ω resistor
- One LED

The Schematic

Figure 17-12 shows the schematic. The pin numbering on the MCP4162 starts at the top left of the package. Pin 1 is indicated by the indented dot to the left of the Microchip logo on the IC (see Figure 17-11).

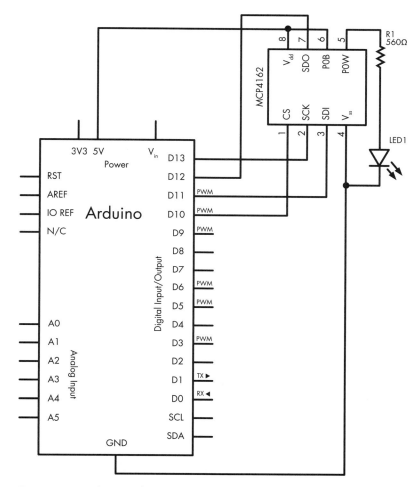

Figure 17-12: Schematic for Project 56

The Sketch

Enter and upload the following sketch:

```
// Project 56 - Using a Digital Rheostat

❶ #include "SPI.h" // necessary library
int ss=10;      // using digital pin 10 for SPI slave select
int del=200;    // used for delaying the steps between LED brightness values

void setup()
{
❷   SPI.begin();
    pinMode(ss, OUTPUT);    // we use this for the SS pin
    digitalWrite(ss, HIGH); // the SS pin is active low, so set it up high first
```

```
❸   SPI.setBitOrder(MSBFIRST);
    // our MCP4162 requires data to be sent MSB (most significant byte) first

  }

❹ void setValue(int value)
  {
    digitalWrite(ss, LOW);
   SPI.transfer(0); // send the command byte
    SPI.transfer(value); // send the value (0 to 255)
    digitalWrite(ss, HIGH);
  }

  void loop()
  {
❺   for (int a=0; a<256; a++)
    {
      setValue(a);
      delay(del);
    }
❻   for (int a=255; a>=0; a--)
    {
      setValue(a);
      delay(del);
    }
  }
```

Let's walk through the code. First, we set up the SPI bus at ❶ and ❷. At ❸, we set the byte direction to suit the MPC4162. To make setting the resistance easier, we use the custom function at ❹, which accepts the resistance step (0 through 255) and passes it to the MCP4162. Finally, the sketch uses two loops to move the rheostat through all the stages, from zero to the maximum at ❺ and then back to zero at ❻. This last piece should make the LED increase and decrease in brightness, fading up and down for as long as the sketch is running.

Looking Ahead

In this chapter you learned about and experimented with two important Arduino communication methods. Now you're ready to interface your Arduino with a huge variety of sensors, more advanced components, and other items as they become available on the market. One of the most popular components today is a real-time clock IC that allows your projects to keep and work with time—and that's the topic of Chapter 18. So let's go!

18

REAL-TIME CLOCKS

In this chapter you will

- Set and retrieve the time and date from a real-time clock module
- Discover new ways to connect devices to an Arduino
- Create a digital clock
- Build an employee RFID time clock

A *real-time clock (RTC)* IC module is a small timekeeping device that opens all sorts of possibilities for Arduino projects. Once set with the current time and date, an RTC provides accurate time and date data on request.

You'll find many different RTC ICs on the market, some more accurate than others. In this chapter, we'll use the Maxim DS3232; it doesn't require any external circuitry other than a backup battery, and it's incredibly accurate and quite robust in module form. The DS3232 is available as a break-out board from various retailers, including the version from Freetronics (*http://www.freetronics.com/rtc/*) that is shown in Figure 18-1.

Figure 18-1: Real-time clock IC module

Connecting the RTC Module

It's easy to connect the RTC module to an Arduino, because it uses the I²C bus (discussed in Chapter 17). All you need are four wires: GND and VCC go to Arduino GND and 5V, respectively; SDA and SCL go to Arduino A4 and A5, respectively. Due to the module's design, no extra pull-up resistors are required on the I²C bus, unless you have a long cable between the module and your Arduino. If so, solder across the pads on the underside marked "Pullups: SDA and SCL" to enable the built-in I²C pull-up resistors.

For convenience, consider mounting the module on a blank ProtoShield so it can be integrated easily with other hardware for other projects. And make sure you have the backup battery installed, or your time data will be lost when you turn off the project!

Project #57: Adding and Displaying Time and Date with an RTC

In this project you'll learn how to set the time and date on the RTC and then retrieve and display it in the Serial Monitor. Time and date information can be useful for various types of projects, such as temperature loggers and alarm clocks.

The Hardware

Here's what you'll need to create this project:

- Arduino and USB cable
- Various connecting wires
- Maxim DS3232 RTC module

The Sketch

Connect the module to the Arduino as described earlier in the chapter, and then enter but *do not upload* the following sketch:

```
// Project 57 - Adding and Displaying Time and Date with an RTC
❶ #include "Wire.h"
#define DS3232_I2C_ADDRESS 0x68

// Convert normal decimal numbers to binary coded decimal
❷ byte decToBcd(byte val)
{
  return( (val/10*16) + (val%10) );
}

// Convert binary coded decimal to normal decimal numbers
byte bcdToDec(byte val)
{
  return( (val/16*10) + (val%16) );
}

void setup()
{
  Wire.begin();
  Serial.begin(9600);

  // set the initial time here:
  // DS3232 seconds, minutes, hours, day, date, month, year
❸  setDS3232time(0, 56, 23, 3, 30, 10, 12);

}

❹ void setDS3232time(byte second, byte minute, byte hour, byte dayOfWeek, byte
dayOfMonth, byte month, byte year)
{
  // sets time and date data to DS3232
  Wire.beginTransmission(DS3232_I2C_ADDRESS);
  Wire.write(0); // set next input to start at the seconds register
  Wire.write(decToBcd(second));     // set seconds
  Wire.write(decToBcd(minute));     // set minutes
  Wire.write(decToBcd(hour));       // set hours
  Wire.write(decToBcd(dayOfWeek));  // set day of week (1=Sunday, 7=Saturday)
  Wire.write(decToBcd(dayOfMonth)); // set date (1 to 31)
  Wire.write(decToBcd(month));      // set month
  Wire.write(decToBcd(year));       // set year (0 to 99)
  Wire.endTransmission();
}

❺ void readDS3232time(byte *second,
byte *minute,
byte *hour,
```

```
                byte *dayOfWeek,
                byte *dayOfMonth,
                byte *month,
                byte *year)
{
  Wire.beginTransmission(DS3232_I2C_ADDRESS);
  Wire.write(0); // set DS3232 register pointer to 00h
  Wire.endTransmission();
  Wire.requestFrom(DS3232_I2C_ADDRESS, 7);

  // request seven bytes of data from DS3232 starting from register 00h
  *second     = bcdToDec(Wire.read() & 0x7f);
  *minute     = bcdToDec(Wire.read());
  *hour       = bcdToDec(Wire.read() & 0x3f);
  *dayOfWeek  = bcdToDec(Wire.read());
  *dayOfMonth = bcdToDec(Wire.read());
  *month      = bcdToDec(Wire.read());
  *year       = bcdToDec(Wire.read());
}

void displayTime()
{
  byte second, minute, hour, dayOfWeek, dayOfMonth, month, year;

  // retrieve data from DS3232
❻ readDS3232time(&second, &minute, &hour, &dayOfWeek, &dayOfMonth, &month,
  &year);

  // send it to the serial monitor
  Serial.print(hour, DEC);
  // convert the byte variable to a decimal number when displayed
  Serial.print(":");
  if (minute<10)
  {
      Serial.print("0");
  }
  Serial.print(minute, DEC);
  Serial.print(":");
  if (second<10)
  {
      Serial.print("0");
  }
  Serial.print(second, DEC);
  Serial.print(" ");
  Serial.print(dayOfMonth, DEC);
  Serial.print("/");
  Serial.print(month, DEC);
  Serial.print("/");
  Serial.print(year, DEC);
  Serial.print("  Day of week: ");
  switch(dayOfWeek){
  case 1:
    Serial.println("Sunday");
    break;
```

```
    case 2:
      Serial.println("Monday");
      break;
    case 3:
      Serial.println("Tuesday");
      break;
    case 4:
      Serial.println("Wednesday");
      break;
    case 5:
      Serial.println("Thursday");
      break;
    case 6:
      Serial.println("Friday");
      break;
    case 7:
      Serial.println("Saturday");
      break;
  }
}

void loop()
{
  displayTime(); // display the real-time clock data on the Serial Monitor,
  delay(1000);   // every second
}
```

How It Works

This sketch might look complex, but it's really not so difficult. At ❶, we import the I²C library and set the bus address of the RTC in the sketch as 0x68. At ❷, two custom functions convert decimal numbers to binary-coded decimal (BCD) values and return those values. We perform these conversions because the DS3232 stores values in BCD format.

At ❸, we use the function setDS3232time to pass the time and date information to the RTC IC like this:

```
setDS3232time(second, minute, hour, dayOfWeek, dayOfMonth, month, year)
```

To use this function, simply insert the required data into the various parameters. The *dayOfWeek* parameter is a number between 1 and 7 representing Sunday through Saturday, respectively. The information for *year* is only two digits—for example, you'd use 13 for the year 2013. (The 20 is assumed.) You can insert either fixed values (as in this sketch) or byte variables that contain the parameters.

Thus, to set the time in the RTC, we enter the current date and time values into the setDS3232time function at ❹. Now you can upload the sketch. Having done that once, we comment out the function by placing // in front of the setDS3232time function at ❹, and then we re-upload the sketch to ensure that the time isn't set to the original value every time the sketch starts!

Finally, the function readDS3232time at ❺ reads the time and date from the RTC and inserts the data into byte variables. This is used at ❻ inside the function displayTime, which simply retrieves the data and displays it in the Serial Monitor by printing the contents of the time variables.

Now upload your sketch and open the Serial Monitor. The results should look similar to those shown in Figure 18-2.

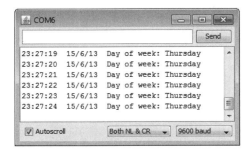

Figure 18-2: Results from Project 57

You can use the contents of the sketch for Project 57 as a basis for other time-related projects. The functions decToBcd, bcdToDec, readDS3232time, and setDS3232time can be inserted and thus reused in future projects. That's one of the benefits of using the Arduino platform: Once you write a useful procedure, it can often be reused later with little or no modification.

Project #58: Creating a Simple Digital Clock

In this project we'll use the functions from Project 57 to display the time and date on a standard character LCD, similar to the one used in the GPS receiver in Project 44.

The Hardware

Here's what you'll need to create this project:

- Arduino and USB cable
- Various connecting wires
- One breadboard
- ProtoScrewShield or similar product
- LCD module or Freetronics LCD shield
- Real-time clock module (shown earlier in the chapter)

First, re-create the hardware used in Project 57. If you connected the RTC module with wires into the Arduino, use a ProtoScrewShield instead to interface with the RTC. Then insert your LCD shield on top of the other shields.

The Sketch

Enter but *do not upload* the following sketch:

```
// Project 58 - Creating a Simple Digital Clock

#include "Wire.h"
#define DS3232_I2C_ADDRESS 0x68

❶ #include <LiquidCrystal.h>
LiquidCrystal lcd( 8, 9, 4, 5, 6, 7 );

// Convert normal decimal numbers to binary coded decimal
byte decToBcd(byte val)
{
  return( (val/10*16) + (val%10) );
}
```

```
// Convert binary coded decimal to normal decimal numbers
byte bcdToDec(byte val)
{
  return( (val/16*10) + (val%16) );
}

void setup()
{
  Wire.begin();
❷  lcd.begin(16, 2);
  // set the initial time here:
  // DS3232 seconds, minutes, hours, day, date, month, year
❸  //setDS3232time(0, 27, 0, 5, 15, 11, 12);
}

void setDS3232time(byte second, byte minute, byte hour, byte dayOfWeek, byte
dayOfMonth, byte month, byte year)
{
  // sets time and date data to DS3232
  Wire.beginTransmission(DS3232_I2C_ADDRESS);
  Wire.write(0);  // set next input to start at the seconds register
  Wire.write(decToBcd(second));     // set seconds
  Wire.write(decToBcd(minute));     // set minutes
  Wire.write(decToBcd(hour));       // set hours
  Wire.write(decToBcd(dayOfWeek));  // set day of week (1=Sunday, 7=Saturday)
  Wire.write(decToBcd(dayOfMonth)); // set date (1 to 31)
  Wire.write(decToBcd(month));      // set month
  Wire.write(decToBcd(year));       // set year (0 to 99)
  Wire.endTransmission();
}

void readDS3232time(byte *second,
byte *minute,
byte *hour,
byte *dayOfWeek,
byte *dayOfMonth,
byte *month,
byte *year)
{
  Wire.beginTransmission(DS3232_I2C_ADDRESS);
  Wire.write(0); // set DS3232 register pointer to 00h
  Wire.endTransmission();
  Wire.requestFrom(DS3232_I2C_ADDRESS, 7);

  // request seven bytes of data from DS3232 starting from register 00h
  *second     = bcdToDec(Wire.read() & 0x7f);
  *minute     = bcdToDec(Wire.read());
  *hour       = bcdToDec(Wire.read() & 0x3f);
  *dayOfWeek  = bcdToDec(Wire.read());
  *dayOfMonth = bcdToDec(Wire.read());
  *month      = bcdToDec(Wire.read());
  *year       = bcdToDec(Wire.read());
}
```

```
void displayTime()
{
  byte second, minute, hour, dayOfWeek, dayOfMonth, month, year;

// retrieve data from DS3232
  readDS3232time(&second, &minute, &hour, &dayOfWeek, &dayOfMonth, &month,
  &year);

  // send the data to the LCD shield
  lcd.clear();
  lcd.setCursor(4,0);
  lcd.print(hour, DEC);
  lcd.print(":");
  if (minute<10)
  {
    lcd.print("0");
  }
  lcd.print(minute, DEC);
  lcd.print(":");
  if (second<10)
  {
    lcd.print("0");
  }
  lcd.print(second, DEC);

  lcd.setCursor(0,1);
  switch(dayOfWeek){
  case 1:
    lcd.print("Sun");
    break;
  case 2:
    lcd.print("Mon");
    break;
  case 3:
    lcd.print("Tue");
    break;
  case 4:
    lcd.print("Wed");
    break;
  case 5:
    lcd.print("Thu");
    break;
  case 6:
    lcd.print("Fri");
    break;
  case 7:
    lcd.print("Sat");
    break;
  }
  lcd.print(" ");
  lcd.print(dayOfMonth, DEC);
  lcd.print("/");
  lcd.print(month, DEC);
```

```
    lcd.print("/");
    lcd.print(year, DEC);
}

void loop()
{
    displayTime(); // display the real-time clock time on the LCD,
    delay(1000);   // every second
}
```

How It Works and Results

The operation of this sketch is similar to that of Project 57, except in this
case, we've altered the function displayTime to send time and date data to
the LCD instead of to the Serial Monitor, and we've added the setup lines
required for the LCD at ❶ and ❷. (For a refresher on using the LCD module, see Chapter 7.) Don't forget to upload the sketch first with the time and
date data entered at ❸, and then re-upload the sketch with that point com-
mented out. After uploading the sketch, your results should be similar to
those shown in Figure 18-4.

Figure 18-4: Display from Project 58

Now that you've worked through Projects 57 and 58, you should have a
sense of how to read and write data to and from the RTC IC in your sketches.
Now let's use what you've learned so far to create something really useful.

Project #59: Creating an RFID Time-Clock System

In this project we'll create a time-clock system. You'll see how Arduino
shields can work together and how the ProtoScrewShield helps you introduce electronic parts that aren't mounted on a shield. This system can be
used by two people who are assigned an RFID card or tag that they'll swipe
over an RFID reader when they enter or leave an area (such as the work-
place or a home). The time and card details will be recorded to a microSD
card for later analysis.

We covered logging data to a microSD card in Chapter 13, to the RFID
in Chapter 16, and to the RTC earlier in this chapter. Now we'll put the
pieces together.

The Hardware

Here's what you'll need to create this project:

- Arduino and USB cable
- Various connecting wires
- Real-time clock module (shown earlier in the chapter)
- LCD module or Freetronics LCD shield
- MicroSD card shield and card (from Chapter 13)
- ProtoScrewShield or similar product
- RFID reader module and two tags (from Chapter 16)

To assemble the system, start with the Arduino Uno at the bottom, and then add your ProtoScrewShield, the microSD card shield, and the LCD shield on the top. Connect the RFID reader as you did in Chapter 16, and connect the RTC module as described earlier in this chapter. The assembly should look similar to that shown in Figure 18-5.

Figure 18-5: The time-clock assembly

The Sketch

Now enter and upload the following sketch. Remember that when you're uploading sketches to an RFID-connected Arduino, you need to ensure that you remove the wire between the RFID reader RX and Arduino pin D0, and then reconnect it once the sketch has been uploaded successfully.

```
// Project 59 - Creating an RFID Time-Clock System
```

❶
```
#include "Wire.h" // for RTC
#define DS3232_I2C_ADDRESS 0x68
```
❷
```
#include "SD.h"  // for SD card
```

```
#include <LiquidCrystal.h>
LiquidCrystal lcd( 8, 9, 4, 5, 6, 7 );
int data1 = 0;
```

❸
```
// Use Listing 16-1 to find your tag numbers
int Mary[14] = {
  2, 52, 48, 48, 48, 56, 54, 67, 54, 54, 66, 54, 66, 3};
int John[14] = {
  2, 52, 48, 48, 48, 56, 54, 66, 49, 52, 70, 51, 56, 3};
int newtag[14] = {
  0,0,0,0,0,0,0,0,0,0,0,0,0,0}; // used for read comparisons

// Convert normal decimal numbers to binary coded decimal
byte decToBcd(byte val)
{
  return( (val/10*16) + (val%10) );
}

// Convert binary coded decimal to normal decimal numbers
byte bcdToDec(byte val)
{
  return( (val/16*10) + (val%16) );
}

void setDS3232time(byte second, byte minute, byte hour, byte dayOfWeek, byte
dayOfMonth, byte month, byte year)
{
  // Sets time and date data to DS3232
  Wire.beginTransmission(DS3232_I2C_ADDRESS);
  Wire.write(0);  // set next input to start at the seconds register
  Wire.write(decToBcd(second));      // set seconds
  Wire.write(decToBcd(minute));      // set minutes
  Wire.write(decToBcd(hour));        // set hours
  Wire.write(decToBcd(dayOfWeek));   // set day of week (1=Sunday, 7=Saturday)
  Wire.write(decToBcd(dayOfMonth));  // set date (1 to 31)
  Wire.write(decToBcd(month));       // set month
  Wire.write(decToBcd(year));        // set year (0 to 99)
  Wire.endTransmission();
}

void readDS3232time(byte *second,
byte *minute,
byte *hour,
byte *dayOfWeek,
byte *dayOfMonth,
byte *month,
byte *year)
```

```
{
  Wire.beginTransmission(DS3232_I2C_ADDRESS);
  Wire.write(0); // set DS3232 register pointer to 00h
  Wire.endTransmission();
  Wire.requestFrom(DS3232_I2C_ADDRESS, 7);

  // Request seven bytes of data from DS3232 starting from register 00h
  *second     = bcdToDec(Wire.read() & 0x7f);
  *minute     = bcdToDec(Wire.read());
  *hour       = bcdToDec(Wire.read() & 0x3f);
  *dayOfWeek  = bcdToDec(Wire.read());
  *dayOfMonth = bcdToDec(Wire.read());
  *month      = bcdToDec(Wire.read());
  *year       = bcdToDec(Wire.read());
}

void setup()
{
  Serial.flush(); // need to flush serial buffer
  Serial.begin(9600);
  Wire.begin();
  lcd.begin(16, 2);
  // set the initial time here:
  // DS3232 seconds, minutes, hours, day, date, month, year
  //setDS3232time(0, 27, 0, 5, 15, 11, 12);

  // Check that the microSD card exists and can be used
  if (!SD.begin(8))
  {
    lcd.print("uSD card failure");
    // stop the sketch
    return;
  }
  lcd.print("uSD card OK");
  delay(1000);
  lcd.clear();
}

// Compares two arrays and returns true if identical.
// This is good for comparing tags.
boolean comparetag(int aa[14], int bb[14])
{
  boolean ff=false;
  int fg=0;
  for (int cc=0; cc<14; cc++)
  {
    if (aa[cc]==bb[cc])
    {
      fg++;
    }
  }
  if (fg==14)
  {
    ff=true;      // all 14 elements in the array match each other
  }
```

❹

```
    return ff;
}

void wipeNewTag()
{
  for (int i=0; i<=14; i++)
  {
    newtag[i]=0;
  }
}

void loop()
{
  byte second, minute, hour, dayOfWeek, dayOfMonth, month, year;

  if (Serial.available() > 0) // if a read has been attempted
  {
    // Read the incoming number on serial RX
    delay(100);  // Allow time for the data to come in from the serial buffer
    for (int z=0; z<14; z++) // read the rest of the tag
    {
      data1=Serial.read();
      newtag[z]=data1;
    }
    Serial.flush(); // stops multiple reads
    // retrieve data from DS3232
    readDS3232time(&second, &minute, &hour, &dayOfWeek, &dayOfMonth, &month,
    &year);
  }

  //now do something based on the tag type
❺  if (comparetag(newtag, Mary) == true)
  {
    lcd.print("Hello Mary ");
    File dataFile = SD.open("DATA.TXT", FILE_WRITE);
    if (dataFile)
    {
      dataFile.print("Mary ");
      dataFile.print(hour);
      dataFile.print(":");
      if (minute<10) { dataFile.print("0"); }
      dataFile.print(minute);
      dataFile.print(":");
      if (second<10) { dataFile.print("0"); }
      dataFile.print(second);
      dataFile.print(" ");
      dataFile.print(dayOfMonth);
      dataFile.print("/");
      dataFile.print(month);
      dataFile.print("/");
      dataFile.print(year);
      dataFile.println();
      dataFile.close();
    }
```

```
    delay(1000);
    lcd.clear();
    wipeNewTag();
  }

  if (comparetag(newtag, John)==true)
  {
    lcd.print("Hello John ");
    File dataFile = SD.open("DATA.TXT", FILE_WRITE);
    if (dataFile)
    {
      dataFile.print("John ");
      dataFile.print(hour);
      dataFile.print(":");
      if (minute<10) { dataFile.print("0"); }
      dataFile.print(minute);
      dataFile.print(":");
      if (second<10) { dataFile.print("0"); }
      dataFile.print(second);
      dataFile.print(" ");
      dataFile.print(dayOfMonth);
      dataFile.print("/");
      dataFile.print(month);
      dataFile.print("/");
      dataFile.print(year);
      dataFile.println();
      dataFile.close();
    }
    delay(1000);
    lcd.clear();
    wipeNewTag();
  }
}
```

How It Works

In this sketch, the system first waits for an RFID card to be presented to the reader. If the RFID card is recognized, then the card owner's name, the time, and the date are appended to a text file stored on the microSD card.

At ❶ are the functions required for the I^{2}C bus and the real-time clock, and at ❷ is the line required to set up the microSD card shield. At ❹, we check and report on the status of the microSD card. At ❺, the card just read is compared against the stored card number for one of two people—in this case, John and Mary. If there is a match, the data is written to the microSD card. With some modification, you could add more cards to the system simply by adding the card serial numbers below the existing numbers at ❸ and then adding other comparison functions like those at ❺.

When the time comes to review the logged data, simply copy the file *data.txt* from the microSD card and view the data with a text editor; or import it into a spreadsheet for further analysis. The data is laid out so that it's easy to read, as shown in Figure 18-6.

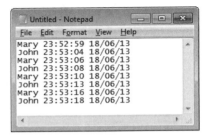

Figure 18-6: Example data generated by Project 59

Looking Ahead

In this chapter you learned how to work with time and date data via the RTC IC. The RFID system described in Project 59 gives you the framework you need to create your own access systems or even track when your children arrive home. In the final two chapters, we'll create projects that will use the Arduino to communicate over the Internet and a cellular phone network.

19

THE INTERNET

In this chapter you will

- Build a web server to display data on a web page
- Use your Arduino to send tweets on Twitter
- Remotely control Arduino digital outputs from a web browser

In this chapter you'll learn how to connect your Arduino to the outside world via the Internet. By doing this, you can broadcast data from your Arduino and remotely control your Arduino from a web browser.

What You'll Need

To build these Internet-related projects, you will need some common hardware, cable, and information.

Let's start with the hardware. You'll need an Ethernet shield with the W5100 controller chip. You have two options to consider: You can use the genuine Arduino-brand Ethernet shield, as shown in Figure 19-1, or you can use an Arduino Uno–compatible board with integrated Ethernet hardware, such as the Freetronics EtherTen shown in Figure 19-2. The latter is a good choice for new projects or those for which you want to save physical space and money. As you can see, the EtherTen has the connectors for Arduino shields, a USB port, an Ethernet socket, and a microSD card socket.

Figure 19-1: Arduino Ethernet shield

Figure 19-2: Freetronics EtherTen

Regardless of your choice of hardware, you'll also need a standard 10/100 CAT5, CAT5E, or CAT6 network cable to connect your Ethernet shield to your network router or Internet modem.

In addition, you'll need the IP address of your network's router gateway or modem, which should look something like this: 192.168.0.1. You'll also need your computer's IP address in the same format as your router's IP address.

Finally, if you want to communicate with your Arduino from outside your home or local area network, you'll need a static, public IP address. A static IP address is a fixed address assigned to your physical Internet connection by your Internet service provider (ISP). Your Internet connection may not have a static IP address by default; contact your ISP to have this activated if necessary. If your ISP cannot offer a static IP or if it costs too much, you can get one through a third-party company, such as no-ip (*http://www.noip.com/*) or Dyn (*http://dyn.com/dns/*). They can set you up with a web address that does not change and will divert users to your current IP address.

Now let's put our hardware to the test with a simple project.

Project #60: Building a Remote-Monitoring Station

In projects in previous chapters, we gathered data from sensors to measure temperature and light. In this project, you'll learn how to display those values on a simple web page that you can access from almost any web-enabled device. This project will display the values of the analog input pins and the status of digital inputs zero to nine on a simple web page as the basis for a remote-monitoring situation.

Using this framework, you can add sensors with analog and digital outputs such as temperature, light, and switch sensors and then display the sensors' status on a web page.

The Hardware

Here's what you'll need to create this project:

- One USB cable
- One network cable
- One Arduino Uno and Ethernet shield, or one Freetronics EtherTen

The Sketch

Enter the following sketch, but *don't upload it* yet:

```
/* Project 60 - Building a Remote-Monitoring Station
   created 18 Dec 2009 by David A. Mellis, modified 9 Apr 2012 by Tom Igoe
   modified 20 Mar 2013 by John Boxall
*/

#include <SPI.h>
#include <Ethernet.h>

IPAddress ip(xxx,xxx,xxx,xxx); //  Replace this with your project's IP address
byte mac[] = { 0xDE, 0xAD, 0xBE, 0xEF, 0xFE, 0xED };
EthernetServer server(80);

void setup()
{
  // Start the Ethernet connection and server
  Ethernet.begin(mac, ip);
  server.begin();
```

❶ IPAddress line
❷ byte mac[] line

```
      for (int z=0; z<10; z++)
      {
        pinMode(z, INPUT); // set digital pins 0 to 9 to inputs
      }
    }

    void loop()
    {
      // listen for incoming clients (incoming web page request connections)
      EthernetClient client = server.available();
      if (client) {
        // an http request ends with a blank line
        boolean currentLineIsBlank = true;
        while (client.connected()) {
          if (client.available()) {
            char c = client.read();
            if (c == '\n' && currentLineIsBlank) {
              client.println("HTTP/1.1 200 OK");
              client.println("Content-Type: text/html");
              client.println("Connection: close");
              client.println();
              client.println("<!DOCTYPE HTML>");
              client.println("<html>");
              // add a meta refresh tag, so the browser pulls again every 5 sec:
❸           client.println("<meta http-equiv=\"refresh\" content=\"5\">");
              // output the value of each analog input pin onto the web page
              for (int analogChannel = 0; analogChannel < 6; analogChannel++) {
                int sensorReading = analogRead(analogChannel);
❹             client.print("analog input ");
                client.print(analogChannel);
                client.print(" is ");
                client.print(sensorReading);
                client.println("<br />");
              }
              // output the value of digital pins 0 to 9 onto the web page
              for (int digitalChannel = 0; digitalChannel < 10; digitalChannel++)
      {

                boolean pinStatus = digitalRead(digitalChannel);
                client.print("digital pin ");
                client.print(digitalChannel);
                client.print(" is ");
                client.print(pinStatus);
                client.println("<br />");
              }
              client.println("</html>");
              break;
            }
            if (c == '\n') {
              // you're starting a new line
              currentLineIsBlank = true;
            }
```

```
      else if (c != '\r') {
        // you've gotten a character on the current line
        currentLineIsBlank = false;
      }
    }
  }
  // give the web browser time to receive the data
  delay(1);
  // close the connection:
  client.stop();
  }
}
```

We'll discuss this sketch in more detail a bit later. First, before upload-ing the sketch, you'll need to enter an IP address for your Ethernet shield so that it can be found on your local network or modem. You can determine the first three parts of the address by checking your router's IP address. For example, if your router's address is 192.168.0.1, change the last digit to something random and different from that of other devices on your net-work, using a number between 1 and 254 that isn't already in use on your network. Enter this at ❶ in the sketch, like so:

```
IPAddress ip(192, 168, 0, 69); // Ethernet shield's IP address
```

Once you've made that change, save and upload your sketch. Next, insert the Ethernet shield into your Arduino if required, connect the network cable into your router or modem and the Ethernet connector, and power on your Arduino board.

Wait about 20 seconds, and then using a web browser on any device or computer on your network, enter the IP address from ❶. If you see something like Figure 19-3, the framework of your moni-toring station is working correctly.

Figure 19-3: Values of the pins mon-itored by our station shown as a web page on any web-connected device with a web browser

Troubleshooting

If this project doesn't work for you, try the following:

- Check that the IP address is set correctly in the sketch at ❶.

- Check that the sketch is correct and uploaded to your Arduino.

- Double-check the local network. You might check to see if a connected computer can access the Internet. If so, check that the Arduino board has power and is connected to the router or modem.

- If you're accessing the project web page from a smartphone, make sure your smartphone is accessing your local Wi-Fi network and not the cell phone company's cellular network.

- If none of the Ethernet shield's LEDs are blinking when the Arduino has power and the Ethernet cable is connected to the shield and router or modem, try another patch lead.

How It Works

If and when your monitoring station is working, you can return to the most important parts of the sketch. The code from the beginning until ❸ is required because it activates the Ethernet hardware, loads the necessary libraries, and starts the Ethernet hardware in void setup. Prior to ❸, the client.print statements are where the sketch sets up the web page to allow it to be read by the web browser. From ❸ on, you can use the functions client.print and client.println to display information on the web page as you would with the Serial Monitor. For example, the following code is used to display the first six lines of the web page shown in Figure 19-3.

```
client.print("analog input ");
client.print(analogChannel);
client.print(" is ");
client.print(sensorReading);
```

At ❹ in the sketch, you see an example of writing text and the contents of a variable to the web page. Here you can use HTML to control the look of your displayed web page, as long as you don't overtax your Arduino's memory. In other words, you can use as much HTML code as you like until you reach the maximum sketch size, which is dictated by the amount of memory in your Arduino board. (The sizes for each board type are described in Table 11-2 on page 218.)

One thing to notice is the MAC address that networks can use to detect individual pieces of hardware connected to the network. Each piece of hardware on a network has a unique MAC address, which can be changed by altering one of the hexadecimal values at ❷. If two or more Arduino-based projects are using one network, you must enter a different MAC address for each device at ❷.

Finally, if you want to view your web page from a device that is not connected to your local network, such as a tablet or phone using a cellular connection, then you'll need to use a technique called *port forwarding* in your network router or modem, with the public IP set up from an organization such as no-ip (*http://www.no-ip.com/*) or Dyn (*http://dyn.com/dns/*). Port forwarding is often unique to the make and model of your router, so do an Internet search for *router port forwarding* or visit a tutorial site such as *http://www.wikihow.com/Port-Forward/* for more information.

Now that you know how to broadcast text and variables over a web page, let's use the Arduino to tweet.

Project #61: Creating an Arduino Tweeter

In this project, you'll learn how to make your Arduino send tweets through Twitter. You can receive all sorts of information that can be generated by a sketch from any device that can access Twitter. If, for example, you want hourly temperature updates from home while you're abroad or even notifications when the kids come home, this can offer an inexpensive solution.

Your Arduino will need its own Twitter account, so do the following:

1. Visit *http://twitter.com/* and create your Arduino's Twitter account. Make note of the username and password.

2. Get a "token" from the third-party handler website *http://arduino-tweet .appspot.com/*, which creates a bridge between your Arduino and the Twitter service. You'll need to follow only step 1 on this site.

3. Copy and paste the token (along with your Arduino's new Twitter account details) into a text file on your computer.

4. Download and install the Twitter Arduino library from *http://playground .arduino.cc/Code/TwitterLibrary/*.

The Hardware

Here's what you'll need to create this project:

- One USB cable
- One network cable
- One Arduino Uno and Ethernet shield, or one Freetronics EtherTen

The Sketch

Enter the following sketch, but *don't upload it* yet:

```
// Project 61 - Creating an Arduino Tweeter
#include <SPI.h>
#include <Ethernet.h>
#include <Twitter.h>
```

❶ `byte ip[] = { xxx,xxx,xxx,xxx };`
❷ `byte mac[] = { 0xDE, 0xAD, 0xBE, 0xEF, 0xFE, 0xED };`
❸ `Twitter twitter("token");`

❹ `char msg[] = "I'm alive!"; // message to tweet`

```
void setup()
{
  delay(30000);
  Ethernet.begin(mac, ip);
  Serial.begin(9600);
}
```

```
void loop()
{
  Serial.println("connecting ...");
  if (twitter.post(msg)) {
    int status = twitter.wait();
    if (status == 200) {
      Serial.println("OK.");
    } else {
      Serial.print("failed : code ");
      Serial.println(status);
    }
  } else {
    Serial.println("connection failed.");
  }
  do {} while (1);
}
```

As with Project 60, insert the IP address at ❶ and modify the MAC address if necessary at ❷. Then insert the Twitter token between the double quotes at ❸. Finally, insert the text that you want to tweet at ❹. Now upload the sketch and connect your hardware to the network. (Don't forget to follow your Arduino's Twitter account with your own account!) After a minute or so, visit your Twitter page and the message should be displayed, as shown in Figure 19-4.

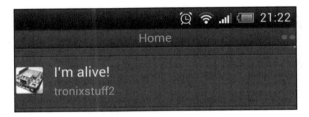

Figure 19-4: Your Arduino's tweet

When you're creating your Arduino tweeter, keep in mind that you can send no more than one tweet per minute and that each message must be unique. (These are Twitter's rules.) When sending tweets, Twitter also replies with a status code. The sketch will receive and display this in the Serial Monitor using the code at ❺, an example of which is shown in Figure 19-5. If you receive a "403" message like this, either your token is incorrect or you're sending tweets too quickly. (For a complete list of Twitter error codes, see *http://dev.twitter.com/docs/error-codes-responses/*.)

Controlling Your Arduino from the Web

You can control your Arduino from a web browser in several different ways. After doing some research, I've found a method that is reliable, secure, and free: Teleduino.

Teleduino is a free service created by New Zealand Arduino enthusiast Nathan Kennedy. It's a simple yet powerful tool for interacting with an Arduino over the Internet. It doesn't require any special or customized Arduino sketches; instead, you simply enter a special URL into a web browser to control the Arduino. You can use Teleduino to control digital output pins and servos, or to send I^{2}C commands, and more features are being

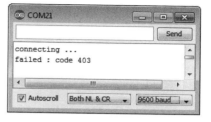

Figure 19-5: Results of tweeting too quickly

added all the time. In Project 62, you'll learn how to configure Teleduino and remotely control digital outputs from a web-enabled device.

Project #62: Setting Up a Remote Control for Your Arduino

Before starting your first Teleduino project, you must register with the Teleduino service and obtain a unique key to identify your Arduino. To do so, visit *https://www.teleduino.org/tools/request-key/* and enter the required information. You should receive an email with your key, which will look something like this: 187654321Z9AEFF952ABCDEF8534B2BBF.

Next, convert your key into an array variable by visiting *https://www.teleduino.org/tools/arduino-sketch-key/*. Enter your key, and the page should return an array similar to that shown in Figure 19-6.

```
byte key[] = { 0x65, 0x9A, 0xCE, 0xB1,
               0xA7, 0x5E, 0x57, 0x2B,
               0x8F, 0xFE, 0xA4, 0x40,
               0x9E, 0x7B, 0x7A, 0xBC };
```

Figure 19-6: A Teleduino key as an array

Each key is unique to a single Arduino, but you can get more keys if you want to run more than one Teleduino project at a time.

The Hardware

Here's what you'll need to create this project:

- One USB cable
- One network cable
- One Arduino Uno and Ethernet shield, or one Freetronics EtherTen
- One 560 Ω resistor (R1)
- One breadboard
- One LED of any color

Assemble your hardware and connect an LED to digital pin 8, as shown in Figure 19-7.

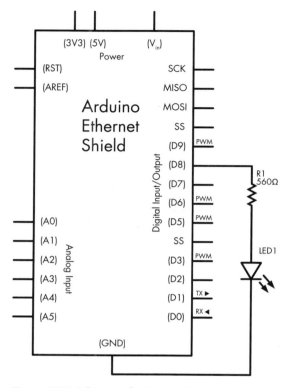

Figure 19-7: Schematic for Project 62

The Sketch

Teleduino projects use only one sketch, which is included with the Teleduino library. Here's how to access the sketch:

1. Download and install the Teleduino library from *https://www.teleduino.org/downloads/*.

2. Restart the Arduino IDE and select **File ▸ Examples ▸ Teleduino328 ▸ TeleduinoEthernetClientProxy**.

3. You should now see the Teleduino sketch. Before uploading it to your Arduino, replace the default key with your key array. The variable you need to replace should be on line 36 of the sketch. Once you've replaced it, save the sketch, and then upload it to your Arduino.

Now connect your hardware to the network and watch the LED. After a minute or so, it should blink a few times and then rest. The number of blinks represents the status of the Teleduino service, as shown in Table 19-1.

Table 19-1: Teleduino Status Blink Code

Number of blinks	Message
1	Initializing
2	Starting network connection
3	Connecting to the Teleduino server
4	Authentication successful
5	Session already exists
6	Invalid or unauthorized key
10	Connection dropped

If you see five blinks, then another Arduino is already programmed with your key and connected to the Teleduino server. At 10 blinks, you should check your hardware and Internet connections. Once the Arduino has connected, it should blink once every 5 seconds or so. Because the status LED is controlled by digital pin 8, you can't use that pin for any other purpose while you're using Teleduino.

Controlling Your Arduino Remotely

To control your Teleduino remotely, you can use any device with a web browser. The command to control the Arduino is sent by entering a URL that you create: *http://us01.proxy.teleduino.org/api/1.0/328.php?k=<YOURKEY> &r=setDigitalOutput&pin=<X>&output=<S>*.

You'll need to change three parameters in the URL: First, replace *<YOURKEY>* with the long alphanumeric key you received from the Teleduino site. Next, replace *<X>* with the digital pin number you want to control. Third, change the *<S>* to 0 for LOW or 1 for HIGH to alter the digital output. For example, to turn digital pin 7 to HIGH, you would enter: *http:// us01.proxy.teleduino.org/api/1.0/328.php?k=<YOURKEY>&r=setDigitalOutput& pin=7&output=1*.

After the command succeeds, you should see something like the following in the web browser:

```
{"status":200,"message":"OK","response"
{"result":0,"time":0.22814512252808,"values":[]}}
```

If the command fails, you should see an error message like this:

```
{"status":403,"message":"Key is offline or invalid.","response":[]}
```

You can send commands to change the digital pins HIGH or LOW by modifying the URL. After you have created the URLs for your project, bookmark them in your browser or create a local web page with the required links as buttons. For example, you might have a URL bookmarked to set digital pin 7 to HIGH and another bookmarked to set it back to LOW.

In some situations, the status of your Arduino outputs could be critical. As a fail-safe in case your Arduino resets itself due to a power outage or other interruption, set the default state for the digital pins. With your project connected to the Teleduino service, visit *https://www.teleduino.org/tools/manage-presets/*. After entering your unique key, you should see a screen of options that allow you to select the mode and value for the digital pins, as shown in Figure 19-8.

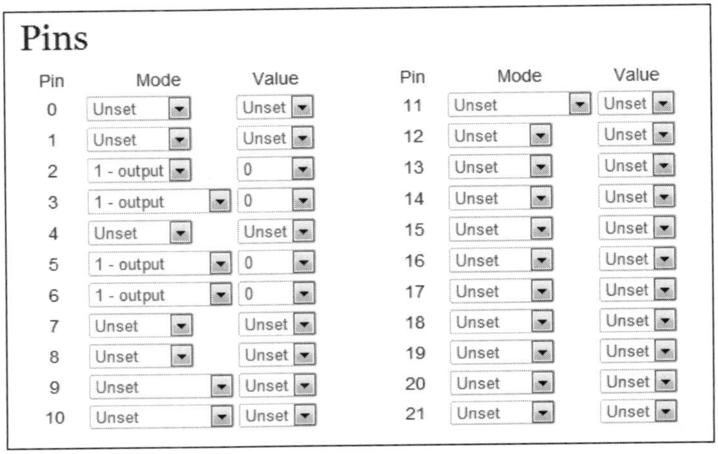

Pins

Pin	Mode	Value	Pin	Mode	Value
0	Unset	Unset	11	Unset	Unset
1	Unset	Unset	12	Unset	Unset
2	1 - output	0	13	Unset	Unset
3	1 - output	0	14	Unset	Unset
4	Unset	Unset	15	Unset	Unset
5	1 - output	0	16	Unset	Unset
6	1 - output	0	17	Unset	Unset
7	Unset	Unset	18	Unset	Unset
8	Unset	Unset	19	Unset	Unset
9	Unset	Unset	20	Unset	Unset
10	Unset	Unset	21	Unset	Unset

Figure 19-8: Default pin status setup page

Looking Ahead

Along with easily monitoring your Arduino over the Internet and having it send tweets on Twitter, you can control your Arduino projects over the Internet without creating any complex sketches, having much networking knowledge, or incurring monthly expenses. By using remote control over the Web, you can control the Arduino from almost anywhere and extend the reach of your Arduino's ability to send data. The three projects in this chapter provide a framework with which you can build upon and design your own remote control projects.

The next chapter, which is the last one in the book, shows you how to make your Arduino send and receive commands over a cellular network connection.

20

CELLULAR COMMUNICATIONS

In this chapter you will

- Have your Arduino dial a telephone number when an event occurs
- Send a text message to a cell phone using the Arduino
- Control devices connected to an Arduino via text message from a cell phone

You can connect your Arduino projects to a cell phone network to allow simple communication between your Arduino and a cellular or landline phone. With a little imagination, you can come up with many uses for this type of communication, including some of the projects included in this chapter.

Be sure to review this chapter before you purchase any hardware, because the success of the projects will depend on your cellular network. Your network must be able to

- Operate at GSM 850 MHz, GSM 900 MHz, DCS 1800 MHz, or PCS 1900 MHz
- Allow the use of devices not supplied by the network provider

Cell networks operating in the European Union, Australia, and New Zealand can usually accommodate these requirements. If you're in the United States and Canada, call your cell provider to ensure that these requirements can be met before you commit to the hardware.

To make use of these projects, you might consider selecting either a prepaid calling plan or a plan that offers a lot of included text messages, in case an error in your sketch causes the project to send out several SMS (Short Message Service) text messages. Also, make sure the requirement to enter a PIN to use the SIM card is turned off. (You should be able to do this easily by inserting the SIM card in a regular cell phone and changing the setting in the security menu.)

The Hardware

Because of hardware restrictions, the projects in this chapter will function only with an Arduino Uno or compatible board (that means no Arduino Mega boards). All the projects use a common hardware configuration, so we'll set that up first.

You'll need specific hardware to complete the projects in this chapter, starting with an SM5100 GSM shield and antenna, shown in Figure 20-1. This shield is available from SparkFun and its distributors. (Look for shield part number CEL-09607, stackable header set PRT-10007, and antenna part number CEL-00675.)

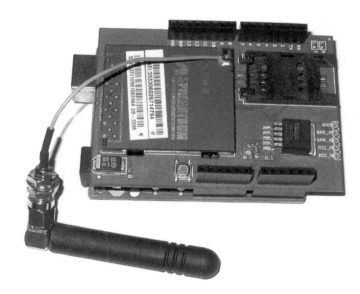

Figure 20-1: GSM shield with antenna attached

You'll also need a power shield and supply. The GSM shield draws up to 2 A of current (more than is available from the Arduino) and will damage your Arduino if it's used without external power. An Arduino-compatible power shield is available from DFRobot at *http://www.dfrobot.com/*, part number DFR0105, and shown in Figure 20-2.

Figure 20-2: DFRobot power shield

Finally, you'll need an external power supply. This can be a DC plug pack or wall wart power supply brick (or a large 7.2 V rechargeable battery, solar panel/battery source, 12 V battery, or similar, as long as it doesn't exceed 35 V DC) that can offer up to 2 A of current.

Preparing the Power Shield

Prepare the power shield, and (this is important) *set its output voltage for use before attaching it to any other parts.* To do this, first remove the jumpers over the top-right and bottom-right header pins (circled on the right side of Figure 20-2). Next, ensure that the two jumpers are set horizontally across the PWRIN pairs at the bottom-left of the shield (circled at the bottom of Figure 20-2).

Now connect your external power supply to the PWRIN terminal block at the bottom-left corner of the power shield. Make sure to match the positive (+) and negative (−) wires correctly. Then turn on the power, and measure the voltage at the PWROUT terminal block at the top-left of the shield while using a small screwdriver to adjust the blue potentiometer, shown in Figure 20-3.

Figure 20-3: Power shield voltage
adjust potentiometer

Continue these adjustments until the voltage measured by the multi-meter from the power shield is 5.0 V DC. When you're done adjusting, turn off the power supply shield.

Now you can put the pieces together:

1. Insert the SIM card into the GSM shield.
2. Plug the GSM shield into the Arduino Uno.
3. Add the power shield on top.
4. Run a small wire from the positive PWROUT terminal block on the power shield to the 5V pin on the shield, and run another wire from the negative PWROUT terminal block to the GND pin on the power shield. These two wires will feed the high-capacity 5 V power from the power shield into the Arduino.

WARNING   *Always ensure that power is applied to the project before connecting the USB cable between the Arduino and the computer. And always remove the USB cable before turning off the external power to the shield.*

Hardware Configuration and Testing

Now let's configure and test the hardware by making sure that the GSM module can communicate with the cellular network and the Arduino. After assembling the hardware and the SIM card, enter and upload the sketch shown in Listing 20-1.

```
// Listing 20-1
// Written by Ryan Owens - SparkFun CC by v3.0 3/8/10

❶ #include <SoftwareSerial.h> // Virtual serial port
❷ SoftwareSerial cell(2,3);
  char incoming_char = 0;
```

```
void setup()
{
  //Initialize serial ports for communication.
  Serial.begin(9600);
❸  cell.begin(9600);
  Serial.println("Starting SM5100B Communication...");
}

void loop()
{
  //If a character comes in from the cellular module...
  if( cell.available() > 0 )
  {
    //Get the character from the cellular serial port.
    incoming_char = cell.read();
    //Print the incoming character to the terminal.
    Serial.print(incoming_char);
  }
  //If a character is coming from the terminal to the Arduino...
  if( Serial.available() > 0 )
  {
    incoming_char = Serial.read();//Get the character coming from the terminal
    cell.print(incoming_char);    //Send the character to the cellular module.
  }
}
```

Listing 20-1: GSM shield test sketch

This sketch simply relays all the information coming from the GSM shield to the Serial Monitor. The GSM shield has a serial connection between it and Arduino digital pins 2 and 3 so that it won't interfere with the normal serial port between the Arduino and the PC, which is on digital pins 0 and 1. We set up a SoftwareSerial virtual serial port for the GSM shield at ❶, ❷, and ❸. (The required library is included with the Arduino IDE.)

Once you've uploaded the sketch, open the Serial Monitor window and wait about 30 seconds. You should see data similar to that shown in Figure 20-4.

You're looking for messages that start with +SIND: because they tell you about the status of the GSM module and its connection with the network. If your messages end with +SIND: 4, your shield has found and connected to the network, and you can move on to Project 63. However, if your messages end with a +SIND: 8, change the frequency that the module uses, as described in the next section.

Figure 20-4: Example output from Listing 20-1

Changing the Operating Frequency

To change your module's operating frequency, follow these steps:

1. Close the Arduino IDE and the Serial Monitor. Then load the terminal software used for Project 48.
2. Select the same COM port used by the Arduino board, and click **Connect**.
3. Look up the frequency used by your cell network in Table 20-1, and note the band number.

Table 20-1: GSM Module Operating Bands

Frequency	Band
GSM 900 MHz	0
DCS 1800 MHz	1
PCS 1900 MHz	2
GSM 850 MHz	3
GSM 900 MHz and DCS 1800 MHz	4
GSM 850 MHz and GSM 900 MHz	5
GSM 850 MHz and DCS 1800 MHz	6
GSM 850 MHz and PCS 1900 MHz	7
GSM 900 MHz and PCS 1900 MHz	8
GSM 850 MHz, GSM 900 MHz, and DCS 1800 MHz	9
GSM 850 MHz, GSM 900 MHz, and PCS 1900 MHz	10

4. Enter the command **AT+SBAND?** and press ENTER. The module should reply with the current band settings, as shown in Figure 20-5.

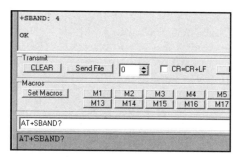

Figure 20-5: Example band interrogation

5. To set the GSM module to the required band, enter the command **AT+SBAND=x**, where **x** is the band number selected from Table 20-1. The shield should return *OK*, as shown in Figure 20-6.

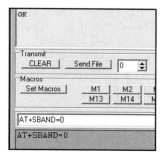

Figure 20-6: Successful band change

6. Finally, reset the Arduino and make sure that it connects to the network and displays +SIND: 4. Connect a jumper wire between the GND and RST pins on the Arduino for a second to trigger a reset, and the status codes should appear on the terminal screen, as shown in Figure 20-7.

```
Starting SM5100B Communication...

+SIND: 1

+SIND: 10,"SM",1,"FD",1,"LD",1,"MC",1,"RC",1,"ME",1

+STIN:0

+SIND: 11

+SIND: 3

+SIND: 4
```

Figure 20-7: Successfully connected to the network

You need to change the SBAND value only once, because the setting is stored in the GSM shield's EEPROM. However, if you relocate and use a different cell network, you may have to change this value.

One final test is needed: Call your GSM shield using another telephone. If you have outgoing caller ID activated on the phone you're calling from, the number should appear in the GSM shield output, as shown in Figure 20-8. (Note that your number will replace the 0452280886 number used in the figure.)

At this stage, you can be confident that everything is working as expected. Now on to the projects!

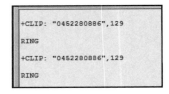

Figure 20-8: Results of calling the GSM shield

Project #63: Building an Arduino Dialer

By the end of this project, your Arduino will dial a telephone number when an event occurs, as determined by your Arduino sketch. For example, if the temperature in your storage freezer rises above a certain level or your burglar alarm system activates, you could have the Arduino call you from a preset number for 20 seconds and then hang up. Your phone's caller ID will identify the phone number as the Arduino.

The Hardware

This project uses the hardware described at the beginning of the chapter as well as any extra circuitry you choose for your application. For demonstration purposes, we'll use a button to trigger the call.

In addition to the hardware already discussed, here's what you'll need to create this project:

- One push button
- One 10 kΩ resistor
- One 100 nF capacitor
- Various connecting wires
- One breadboard

The Schematic

Connect the external circuitry, as shown in Figure 20-9.

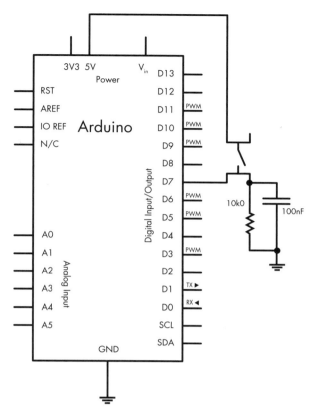

Figure 20-9: Schematic for Project 63

The Sketch

Enter *but don't upload* the following sketch:

```
// Project 63 - Building an Arduino Dialer

#include <SoftwareSerial.h>
SoftwareSerial cell(2,3);

void setup()
{
  pinMode(7, INPUT);
  cell.begin(9600);
  delay(30000); // give the GSM module time to initialize network location
}

void callSomeone()
{
```

❶ `delay(30000); // give the GSM module time to initialize network location`

❷ `void callSomeone()`

```
❸    cell.println("ATDxxxxxxxxxx"); // dial the phone number xxxxxxxxxx
     // change xxxxxxxxxx to your desired phone number (with area code)
     delay(20000); // wait 20 seconds.
     cell.println("ATH"); // end call
     delay(60000); // wait 60 seconds for GSM module
   }

   void loop()
   {
     if (digitalRead(7) == HIGH)
     {
❹      callSomeone();
     }
   }
```

How It Works

The module is activated in void setup, and at ❶ we give it some time to locate and register to the network. For our "event," the Arduino monitors the button connected to digital pin 7. When this button is pressed, the function callSomeone is run at ❷. At ❸, the sketch sends the command to dial a telephone number.

You'll replace *xxxxxxxxxx* with the number you want your Arduino to call. Use the same dialing method that you'd use from your mobile phone. For example, if you wanted the Arduino to call 212.555.1212, you'd add this:

```
cell.println("ATD2125551212");
```

After you have entered the phone number, you can upload the sketch, wait a minute, and then test it by pressing the button. It's very easy to integrate the dialing function into an existing sketch, because it's simply called when required at ❹. From here, it's up to you to find a reason for your Arduino to dial a phone number.

Now let's drag your Arduino into the 21st century by sending a text message.

Project #64: Building an Arduino Texter

In this project, the Arduino will send a text message to another cell phone when an event occurs. To simplify the code, we'll use the SerialGSM Arduino library, available from *https://github.com/meirm/SerialGSM/*. After you've installed the library, restart the Arduino IDE.

The hardware you'll need for this project is identical to that for Project 63.

The Sketch

Enter the following sketch into the Arduino IDE, but *don't upload it* yet:

```
// Project 64 - Building an Arduino Texter

#include <SerialGSM.h>
#include <SoftwareSerial.h>
SerialGSM cell(2,3);                    ❶

void setup()
{
  pinMode(7, INPUT);
  delay(30000); // wait for the GSM module
  cell.begin(9600);
}

void textSomeone()
{
  cell.Verbose(true); // used for debugging
  cell.Boot();
  cell.FwdSMS2Serial();
  cell.Rcpt("+xxxxxxxxxxx");  // replace xxxxxxxxxx with the    ❷
                             // recipient's cell number
  cell.Message("This is the contents of a text message");      ❸
  cell.SendSMS();
}

void loop()
{
  if (digitalRead(7) == HIGH)           ❹
  {
    textSomeone();
  }

  if (cell.ReceiveSMS())
  {
    Serial.println(cell.Message());
    cell.DeleteAllSMS();
  }
}
```

How It Works

The GSM shield is set up as normal at ❶ and in void setup(). Button presses are detected at ❹, and the function textSomeone is called. This simple function sends a text message to the cellular phone number stored at ❷.

Before uploading the sketch, replace *xxxxxxxxxx* with the recipient's cellular phone number in international format: the country code, the area code, and the number, without any spaces or brackets. For example, to send a text to 212.555.1212 in the United States, you would store +12125551212.

The text message to be sent is stored at ❸. (Note that the maximum length for a message is 160 characters.)

After you have stored a sample text message and a destination number, upload the sketch, wait 30 seconds, and then press the button. In a moment, the message should arrive on the destination phone, as shown in Figure 20-10.

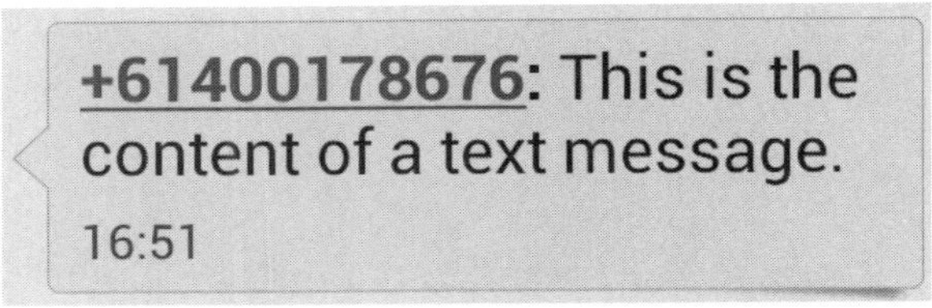

+61400178676: This is the content of a text message.
16:51

Figure 20-10: Sample text message being received

Project 64 can be integrated quite easily into other sketches, and various text messages could be sent by comparing data against a parameter with a switch-case function.

NOTE *Remember that the cost of text messages can add up quickly, so when you're experimenting, be sure that you're using an unlimited or prepaid calling plan.*

Project #65: Setting Up an SMS Remote Control

In this project you'll control the digital output pins on your Arduino by sending a text message from your cell phone. You should be able to use your existing knowledge to add various devices to control. We'll allow for four separate digital outputs, but you can control more or less as required.

To turn on or off four digital outputs (pins 10 through 13 in this example), you'd send a text message to your Arduino in the following format: #a*x*b*x*c*x*d*x*, replacing *x* with either a 0 for off or a 1 for on. For example, to turn on all four outputs, you'd send #a1b1c1d1.

The Hardware

This project uses the hardware described at the start of the chapter, plus any extra circuitry you choose. We'll use four LEDs to indicate the status of the digital outputs being controlled. Therefore, the following extra hardware is required:

- Four LEDs
- Four 560 Ω resistors
- Various connecting wires
- One breadboard

The Schematic

Connect the external circuitry, as shown in Figure 20-11.

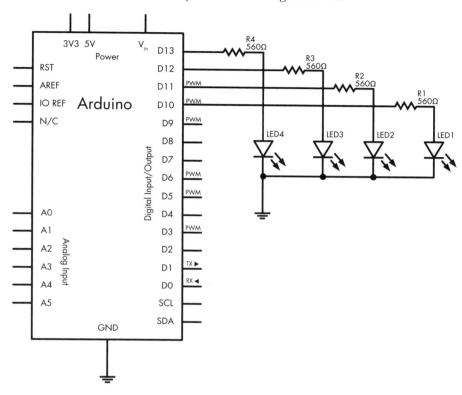

Figure 20-11: Circuitry for Project 65

The Sketch

Enter and upload the following sketch:

```
// Project 65 - Setting Up an SMS Remote Control

#include <SoftwareSerial.h>
SoftwareSerial cell(2,3);
char inchar;

void setup()
{
  // set up digital output pins to control
  pinMode(10, OUTPUT);
  pinMode(11, OUTPUT);
  pinMode(12, OUTPUT);
  pinMode(13, OUTPUT);
```

```
          digitalWrite(10, LOW); // default state for I/O pins at power-up or reset,
          digitalWrite(11, LOW); // change as you wish.
          digitalWrite(12, LOW);
          digitalWrite(13, LOW);

          //Initialize the GSM module serial port for communication.
          cell.begin(9600);
          delay(30000);
❶        cell.println("AT+CMGF=1");
          delay(200);
❷        cell.println("AT+CNMI=3,3,0,0");
          delay(200);
       }

     void loop()
     {
        // If a character comes in from the cellular module...
❸      if(cell.available() > 0)
        {
          inchar = cell.read();
❹        if (inchar == '#') // the start of our command
          {
            delay(10);
            inchar = cell.read();
❺          if (inchar == 'a')
            {
              delay(10);
              inchar = cell.read();
              if (inchar == '0')
              {
                digitalWrite(10, LOW);
              }
              else if (inchar == '1')
              {
                digitalWrite(10, HIGH);
              }
              delay(10);
              inchar = cell.read();
              if (inchar == 'b')
              {
                inchar = cell.read();
                if (inchar == '0')
                {
                  digitalWrite(11, LOW);
                }
                else if (inchar == '1')
                {
                  digitalWrite(11, HIGH);
                }
                delay(10);
                inchar = cell.read();
                if (inchar == 'c')
```

```
    {
      inchar = cell.read();
      if (inchar == '0')
      {
        digitalWrite(12, LOW);
      }
      else if (inchar == '1')
      {
        digitalWrite(12, HIGH);
      }
      delay(10);
      inchar = cell.read();
      if (inchar == 'd')
      {
        delay(10);
        inchar = cell.read();
        if (inchar == '0')
        {
          digitalWrite(13, LOW);
        }
        else if (inchar == '1')
        {
          digitalWrite(13, HIGH);
        }
        delay(10);
      }
    }
    cell.println("AT+CMGD=1,4"); // delete all SMS
  }
      }
    }
  }
}
```

How It Works

In this project the Arduino monitors every text character sent from the GSM module. Thus, at ❶ we tell the GSM shield to convert incoming SMS messages to text and send the contents to the virtual serial port at ❷. Next, the Arduino simply waits for a text message to come from the GSM shield at ❸.

Because the commands sent from the cell phone and passed by the GSM module to control pins on the Arduino start with a #, the sketch waits for a hash mark (#) to appear in the text message at ❹. At ❺, the first output parameter a is checked—if it is followed by a 0 or 1, the pin is turned off or on, respectively. The process repeats for the next three outputs controlled by b, c, and d.

Imagine how easy it would be to use this project to create a remote control for all manner of things, such as lights, pumps, alarms, and more.

Looking Ahead

With the three projects in this chapter, you've created a great framework on which to build your own projects that can communicate over a cell network. You're limited only by your imagination—for example, you could receive a text message if your basement floods or turn on your air conditioner from your cell phone. Once again, remember to take heed of network charges before setting your projects free.

At this point, after having read about (and hopefully built) the 65 projects in this book, you should have the understanding, knowledge, and confidence you need to create your own Arduino-based projects. You know the basic building blocks used to create many projects, and I'm sure you will be able to apply the technology to solve all sorts of problems and have fun at the same time.

I'm happy to discuss the projects in this book and hear your feedback and suggestions via my web forum, located at *http://www.tronixforum.com/?forum=376318.*

But remember—this is only the beginning. You can find many more forms of hardware to work with, and with some thought and planning, you can work with them all. You'll find a huge community of Arduino users on the Internet (in such places as the Arduino forum at *http://arduino.cc/forum/*) and even at a local hackerspace or club.

So don't just sit there—make something!

INDEX

Symbols & Numbers

&, 139
&&, 73
*, 84
*/, 27
==, 71
!, 73
!=, 71
/, 84
/*, 27
//, 27
>, 84
>=, 84
#define, 70
#include, 149
-, 83
<, 84
<=, 84
%, 133
+, 83
|, 139–140
||, 73
1N4004 diode, 51, 232
24LC512. *See* EEPROM
433 MHz receiver shield, 275
7-segment LED displays, 126–128
    controlling with shift registers,
        127–130
    schematic symbol, 127
74HC595. *See* shift registers
7805 voltage regulator, 209
    schematic symbol, 209

A

amperes, 35
analogRead(), 80
analogReference(), 85–86
analogWrite(), 48–49
and, 73
Arduino, 1
    board types, 217–224
    libraries. *See* libraries
    microcontroller specifications,
        217–218
        ATmega2560, 218
        ATmega328P-PU, 218
        ATmega328P SMD, 218
        SAM3X8E, 218
    shields. *See* shields

sketches
    adding comments to, 27
    creating your first, 27
    uploading, 30
    verifying, 30
suppliers, 6
Arduino Due, 223–224
    specifications, 218
Arduino LilyPad, 221–222
Arduino Mega2560, 222
    specifications, 218
Arduino Nano, 221
Arduino Uno, 20
    analog sockets, 22
    boards compatible with, 219–220
    DC socket, 20
    digital I/O sockets, 22
    onboard LED, 22
    power connector, 20
    power sockets, 22
    reset button, 23
    schematic symbol, 57
AREF pin. *See* reference voltage
arithmetic, 83–84
arrays
    defining, 124
    writing to and reading from, 125–126
ATmega2560 specifications, 218
ATmega328P-PU, 21, 210, 211, 213,
        216, 218
    Arduino equivalent pinouts, 213
    pin labels, 214
    schematic symbol, 210
    specifications, 218
    uploading sketches to, 214–217
        microcontroller swap method,
            214–215
        using existing Arduino board,
            215–216
        using FTDI cable, 216–217
ATmega328 SMD, specifications, 218
attachInterrupt(), 185

B

battery tester, 80–83
BC548 transistor, 50
binary numbers, 116
    displaying with LEDs, 119–121
    game, 122–124
binary to base-10 conversion, 116–117

bitwise arithmetic, 139–141
    AND, 139
    bitshift left and right, 140–141
    NOT, 140
    OR, 139–140
    XOR, 140
blinking
    an LED, 29–30
    a wave pattern, 43
Boolean variables, 72
bootloader, 192
breadboard Arduino, 208–217
    circuit schematic, 211
buttons. *See* push buttons
buzzers. *See* piezoelectric buzzers
byte variables, 117

C

capacitors, 60–62
    ceramic, 61
        schematic symbol, 61
    electrolytic, 62
        schematic symbol, 62
    measuring capacity of, 60–61
    reading values of, 61
cellular communications
    controlling Arduino via text message,
        360–363
    making calls from Arduino, 356–358
    SerialGSM library, 358
    sending text messages from Arduino,
        358–360
    SM5100B GSM shield, 350
        changing operating frequency,
            354–355
        configuring and testing, 352–353
        suitable antenna, 350
        suitable power supply, 351
        supported GSM network
            frequencies, 349
client.print(), 342
client.println(), 342
clock. *See* real-time clock
collision detection
    with infrared sensors, 249–251
    with ultrasonic sensors, 251–256
comments, 27
comparison operators, 72–73, 84
    and, 73
    not, 73
    or, 73
compiling sketches, 30–31
constants. *See* #define
crystal oscillators, 209–210
    schematic symbol, 210
current (electrical), 34

D

Darlington transistors, 231–232
    schematic symbol, 231
data buses. *See* I²C bus; SPI bus
#define, 70
delay(), 29
delayMicroseconds(), 252
dice, 113–115
digital clock. *See* real-time clock
digital inputs, 63
    activating, 70
    reading, 70
digital outputs
    activating, 28
    maximum current, 49
    pulse-width modulation, 48
digital rheostats, 318–320
digitalWrite(), 29
diodes, 50–51
    schematic symbol, 57
displaying binary numbers, 119
do while, 105–106
DS3232, 321. *See also* real-time clock

E

EEPROM, 218, 309
    external, 309
    internal Arduino, 301–303
    Microchip 24LC512, 309
EEPROM.h, 302
EEPROM.read, 302
EEPROM.write, 302
electricity
    current, 34
    power, 35
    voltage, 35
electronic components
    capacitors. *See* capacitors
    crystal oscillators, 209–210
        schematic symbol, 210
    Darlington transistors, 231–232
        Schematic symbol, 231
    digital rheostats, 318–320
    diodes, 50–51
        schematic symbol, 57
    DS3232, 321. *See also* real-time clock
    EEPROM. *See* EEPROM
    infrared receivers, 286
    LCDs. *See* liquid crystal displays
        (LCDs)
    LEDs. *See* light-emitting diodes (LEDs)
    microswitches, 243–246
        schematic symbol, 244
    motors, 231–235
        demonstration circuit, 232–235
        stall current, 231

numeric keypad, 187–193
  wiring to Arduino, 188
piezoelectric buzzers, 87
  schematic symbol, 88
port expanders. *See* Microchip
  Technology MCP23017
potentiometer. *See* variable resistors
push buttons. *See* push buttons
relays, 51
  schematic symbol, 58
resistors. *See* resistors
shift registers. *See* shift registers
servos. *See* servos
temperature sensors, 90
  schematic symbol, 91
touchscreens. *See* touchscreens
transistors, 50
  schematic symbol, 58
  switching higher currents with,
    50, 52
trimpots, 87
variable resistors. *See* variable resistors
Eleven, 219
else, 71
EtherMega, 222–223
Ethernet shield, 24, 328
EtherTen, 338

F

false (Boolean value), 72
farads, 61
flash memory, 218
float variables, 84
for, 47
Freeduino, 220
Freetronics
  Eleven, 219
  EtherMega, 222–223
  EtherTen, 338
  RTC module, 322
functions, creating your own, 95–98

G

GLCD.ClearScreen(), 155
GLCD.CursorTo(), 155
GLCD.DrawCircle(), 157
GLCD.DrawHoriLine(), 157
GLCD.DrawRect(), 157
GLCD.DrawRoundRect(), 157
GLCD.DrawVertLine(), 157
GLCD.FillRect(), 157
glcd.h. *See* graphic LCDs
GLCD.init(), 155
GLCD.PrintNumber(), 155
GLCD.Puts(), 155
GLCD.SelectFont(), 155

GLCD.SetDot(), 157
Global Positioning System. *See* GPS
  (Global Positioning System)
GND. *See* ground
Google Maps, 263
GPS (Global Positioning System)
  Arduino shield, 258
  displaying coordinates from, 261–263
  displaying logged journeys on Google
    Maps, 268–269
  displaying time from, 263–265
  logging position data from, 265–267
  receiver, 259
  showing location on Google Maps, 263
  testing GPS shield, 259–261
graphic LCDs. *See also* liquid crystal
  displays (LCDs)
  Arduino library, 155
  connecting to Arduino, 154
  displaying graphics on, 157–160
  displaying text on, 156
  using with Arduino, 153–160
ground, 35
  schematic symbol, 59

H

HD44780. *See* liquid crystal
  displays (LCDs)
higher-voltage circuits, 52–53
HTML, 342

I

I^2C bus
  Arduino connectors, 308
  device address, 308
  EEPROM. *See* EEPROM
  port expanders. *See* Microchip
    Technology MCP23017
  real-time clock. *See* real-time clock
  receiving data, 309
  transmitting data, 308
  voltage warning, 308
IDE. *See* Integrated Development
  Environment (IDE)
if, 71
#include, 149
infrared distance sensor, 246–249
  connection to Arduino, 247
  detecting collisions with, 249–251
  example sketch, 248
infrared remote control
  controlling Arduino with, 289–290
  example Sony codes, 288
  receiver modules, 286
  testing reception, 287–288
  TSOP4138, 286

int, 46
integers, 46
Integrated Development Environment (IDE), 25
  command area, 25
  icons, 25, 26
  installing
    on Mac OS X, 7–10
    on Ubuntu Linux 9.04 and later, 15–18
    on Windows XP and later, 11–15
  menu items, 25, 26
  message window area, 25, 26–27
  text area, 25, 26
  title bar, 5
interrupts, 184–186
  configuring, 184
interrupts(), 185
ip(), 341
IP address, 341

K

keypad.h, 191
keypads. *See* numeric keypads
Knight Rider, 43
KS0066. *See* liquid crystal displays (LCDs)
KS0108B. *See* graphic LCDs; liquid crystal displays (LCDs)

L

lcd.begin(), 150
lcd.clear(), 152
lcd.createChar(), 150
lcd.print(), 150
LCDs. *See* liquid crystal displays (LCDs)
lcd.setCursor(), 150
lcd.write(), 153
LEDs. *See* light-emitting diodes (LEDs)
libraries, 169–173
  installing
    in Mac OS X, 170–171
    in Ubuntu Linux, 172–173
    in Windows XP and later, 171–172
light-emitting diodes (LEDs), 39–40
  Arduino onboard, 22
  calculating current flow, 40
  changing brightness with PWM, 47–49
  matrix modules, 135–137
    animation with, 145–146
    schematic symbols, 136
    using with Arduino, 141–146
  schematic symbol, 57

seven-segment, 126–128
  controlling with shift registers, 127–130
  schematic symbol, 127
  use of, 39
LilyPad Arduino, 221–222
linear, 86
liquid crystal displays (LCDs)
  Arduino library, 150
  character displays, 148
  creating custom characters, 152–153
  graphic. *See* graphic LCDs
  using with Arduino, 149–151
LiquidCrystal LCD(), 150
logarithmic, 86
long variables, 107, 180
loop(), 28

M

MAC address, 342
Maxim DS3232. *See* real-time clocks
Microchip Technology MCP23017, 313–315
Microchip Technology MCP4162, 318–320
micros(), 179–181
MicroSD memory cards, 173–177
  writing data to, 175–177
microswitches, 243–246
  schematic symbol, 244
millis(), 179–181
MISO pin. *See* SPI bus
modulo, 133
MOSI pin. *See* SPI bus
motors, 231–235
  demonstration circuit, 232–235
  stall current, 231
motor shield, 238–240
  connections, 239
multimeters, 38
multiplying numbers, 106–107

N

noInterrupts(), 185
not, 73, 140
numeric keypads, 187–193
  wiring to Arduino, 188

O

ohms, 36
Ohm's Law, 40
or, 73, 139–140
oscilloscopes, 64, 79–80

P

Parallax Ping))). *See* ultrasonic distance sensors
piezoelectric buzzers, 87
    schematic symbol, 88
PIN. *See* numeric keypads
pinMode(), 28
planning your projects, 34
Pololu RP5 tank chassis, 235–242
    controlling with infrared remote, 291–293
port expanders. *See* Microchip Technology MCP23017
potentiometers, 86–87
power (electrical), 35
PowerSwitch Tail, 93
ProtoScrewShield, 327
Pro Trinket, 220–221
pull-down resistors, 65
pulseDuration(), 252
pulse-width modulation (PWM), 47–48
    demonstrating, 49
    output pins, 48
    using with analogWrite(), 48–49
push buttons, 63
    de-bouncing circuit, 66
    pin alignment of, 63
    schematic symbol, 63
    simple example, 65–70
PWM. *See* pulse-width modulation (PWM)

R

radio-frequency identification (RFID)
    125 kHz RFID reader, 296
    controlling Arduino with, 299–301
    defined, 295
    reading tags, 297–298
    time-clock system, 330–336
random(), 112
random numbers, 112
randomSeed(), 112
real-time clock
    connecting to Arduino, 322
    displaying time with, 322–326
    reading the time from, 326
    setting the time, 325
reference voltage, 84–86
    external, 85
    internal, 86
relays, 51
    schematic symbol, 58
repeating functions, 46
resistors, 35–37
    color bands, 36
    power rating, 38
    pull-down, 65

reading resistor values, 36
schematic symbol, 57
variable, 86–87
RF data modules
    Arduino library, 272
    example schematics, 273–274
    wireless remote control with, 272–277
RFID. *See* radio-frequency identification
RTC module, 322

S

safety warning, 18
SAM3X8E, specifications, 218
schematic diagrams, 56
SCK pin. *See* SPI bus
SCL pin. *See* I²C bus
SDA pin. *See* I²C bus
Serial.available(), 107
Serial.begin(), 102
serial buffer, 106
Serial.flush(), 107
SerialGSM.h, 358
serial monitor, 102–103
    debugging with, 105
    displaying text and data in, 102–103
    sending data from serial monitor to Arduino, 106–107
Serial.print(), 102
Serial.println(), 103
servo, 228
    .attach(), 228
    .write(), 228
servo.h, 227
servos, 226–230
    connecting to an Arduino, 227
    example project schematic, 229
    required Arduino functions, 227–228
    selecting an appropriate, 226
setup(), 28
shields, 23
    Ethernet, 24, 328
    microSD card, 163
    numeric display/temperature, 24
    prototyping, 164
        creating your own, 165–169
shiftOut(), 121
shift registers
    74HC595, 118
    clock, 118–119
    data, 118–119
    latch, 118–119
    schematic symbol, 119
sketches
    adding comments to, 27
    creating your first, 27
    uploading, 30
    verifying, 30

SoftwareSerial.h, 353
soldering, 167
soldering iron, 167
solderless breadboard, 41
SPI.begin(), 317
SPI bus, 315
    Arduino connectors, 316
    digital rheostat, 318
    receiving data, 309
    transmitting data, 308
    typical device connection, 316, 317
    voltage warning, 308
SPI.h, 316
SPI.setBitOrder(), 317
SPI.transfer(), 317
SRAM, 218
SS pin. *See* SPI bus
stopwatch, 181–183
suppliers, 6
switch bounce, 64
switch... case, 190

T

tank chassis *See* Pololu RP5 tank chassis
tank robot, 235–242
Teleduino, 344
    blink codes, 347
    controlling Arduino with, 345–348
    default pin status setup, 348
    key, 345
    messages from, 347
    registering with, 345
temperature logging, 177–179
temperature sensor, 90
terminal emulator software, 280
then, 71
thermometer
    analog, 228–230
    digital, 134–135
    monitor, 157–160
    quick-read, 90–92, 99–101
time. *See* real-time clock
timing with Arduino, 179–183, 321
TIP120, 231–232
    schematic symbol, 231
TMP36, 90. *See also* thermometer
    schematic symbol, 91
TO-220, 209, 232
touchscreens, 196–205
    breakout board, 196
    connections to Arduino, 196
    controlling Arduino with, 200
    mapping touchscreen area, 199
traffic light simulator, 74–79
transistors, 50
    schematic symbol, 58
    switching higher currents with, 50, 52

trimpots, 87. *See also* variable resistors
true (Boolean value), 72
TSOP4138, 286
twitter, 343
twitter.h, 343
two-wire interface. *See* I²C bus
types of Arduino, 218

U

ultrasonic distance sensors, 251–256
    connection to Arduino, 252
    detecting collisions with, 254
    example sketch, 252–253
unsigned long variable, 180
uploading sketches
    with FTDI cable, 216–217
    with the IDE, 31

V

variable resistors, 86–87
variables, 45–46
    byte, 117
    float, 84
    integer, 46
    long, 107, 180
verifying sketches, 30–31
void loop(), 28–29
void setup(), 28
virtual serial port, 353
virtualwire.h, 274
voltage (electrical), 35
volts, 35

W

watts, 35
web server, 339
while, 105
Wire.begin(), 308
Wire.beginTransmission(), 309
Wire.endTransmission(), 309
wire.h, 308
wireless data. *See* RF data modules; XBee
Wire.read(), 309
Wire.requestFrom(), 309
Wire.write(), 308

X

XBee, 277
    Arduino shield, 278
    explorer board, 278
    transmitting data with, 279–280
    using a remote control with, 281–283
XOR, 140

The Electronic Frontier Foundation (EFF) is the leading organization defending civil liberties in the digital world. We defend free speech on the Internet, fight illegal surveillance, promote the rights of innovators to develop new digital technologies, and work to ensure that the rights and freedoms we enjoy are enhanced — rather than eroded — as our use of technology grows.

EFF.ORG
ELECTRONIC FRONTIER FOUNDATION
Protecting Rights and Promoting Freedom on the Electronic Frontier